Human Gait - Recent Findings and Research

Edited by Manuel Domínguez-Morales and Francisco Luna-Perejón

Published in London, United Kingdom

Human Gait - Recent Findings and Research
http://dx.doi.org/10.5772/intechopen.111176
Edited by Manuel Domínguez-Morales and Francisco Luna-Perejón

Contributors
Britam Arom Arom Gómez Arias, Cathy Harro, David W. Zeitler, Dorothy Kilvington, Elizaveta S. Koneva, Francisco Luna-Perejón, Galina V. Derevyashkina, Gordon Alderink, Hoda Salsabili, Konstantin V. Lyadov, Larry Duncan, Lauren Hickox, Mackenzi Pergolotti, Manuel Jesus Domínguez-Morales, Miad Faezipour, Pablo Esteban Aqueveque Navarro, Paige Pryson, Paulina Cecilia Ortega-Bastidas, Rebecca Prevost, S. Abdollah Mirbozorgi, Sarah Ostadabbas, Sebastián Gonzalo Chávez Orellana, Shaotong Zhu, Smith Giri, Vitaly G. Polushkin

First published in London, United Kingdom, 2024 by IntechOpen
IntechOpen is the global imprint of INTECHOPEN LIMITED, registered in England and Wales, registration number: 11086078, 167-169 Great Portland Street, London, W1W 5PF, United Kingdom

British Library Cataloguing-in-Publication Data
A catalogue record for this book is available from the British Library

Additional hard and PDF copies can be obtained from orders@intechopen.com

Human Gait - Recent Findings and Research
Edited by Manuel Domínguez-Morales and Francisco Luna-Perejón
p. cm.

This title is part of the Biomedical Engineering Book Series, Volume 24
Topic: Bioinspired Technology and Biomechanics
Series Editor: Robert Koprowski
Topic Editor: Adriano Andrade

Print ISBN 978-1-80356-773-0
Online ISBN 978-1-80356-774-7
eBook (PDF) ISBN 978-1-80356-775-4
ISSN 2631-5343

For EU product safety concerns:
IN TECH d.o.o., Prolaz Marije Krucifikse Kozulić 3, 51000 Rijeka, Croatia,
info@intechopen.com or visit our website at intechopen.com.

IntechOpen Book Series
Biomedical Engineering
Volume 24

Aims and Scope of the Series

Biomedical Engineering is one of the fastest-growing interdisciplinary branches of science and industry. The combination of electronics and computer science with biology and medicine has improved patient diagnosis, reduced rehabilitation time, and helped to facilitate a better quality of life. Nowadays, all medical imaging devices, medical instruments, or new laboratory techniques result from the cooperation of specialists in various fields. The series of Biomedical Engineering books covers such areas of knowledge as chemistry, physics, electronics, medicine, and biology. This series is intended for doctors, engineers, and scientists involved in biomedical engineering or those wanting to start working in this field.

Meet the Series Editor

Robert Koprowski, MD (1997), Ph.D. (2003), Habilitation (2015), is an employee of the University of Silesia, Poland, Institute of Computer Science, Department of Biomedical Computer Systems. For 20 years, he has studied the analysis and processing of biomedical images, emphasizing the full automation of measurement for a large inter-individual variability of patients. Dr. Koprowski has authored more than a hundred research papers with dozens in impact factor (IF) journals and has authored or co-authored six books. Additionally, he is the author of several national and international patents in the field of biomedical devices and imaging. Since 2011, he has been a reviewer of grants and projects (including EU projects) in biomedical engineering.

Meet the Volume Editors

Manuel Domínguez-Morales is a computer engineer with an MSc in Software Engineering and Technology, an MSc in Computer and Networks Engineering, and a Ph.D. in Industrial Informatics. He is an associate professor in the Computer Architecture and Technology Department, University of Seville, Spain. His research focuses on intelligent embedded devices, e-health, computer vision, robotics, signal processing, and AI diagnostic-aid systems. Dr. Domínguez-Morales has more than 120 publications to his credit, including 40 international journal articles. He has also edited four books. He is a journal editor, guest editor, and Program/Academic Committee member in more than 20 international congresses.

Francisco Luna-Perejón is a computer and biomedical engineer with an MSc in Computer Science and a Ph.D. in Computer Science. He is a postdoc lecturer in the Computer Architecture and Technology Department, University of Seville, Spain. His current research interests include deep learning algorithms for diagnostic support and medical instrumentation regarding fall detection systems, gait analysis, and posture recommendations. He has more than forty international publications to his credit, including twenty-two journal articles.

Contents

Preface

The study of human gait, the coordination of muscles, bones, and neural pathways that allows us to walk, run, and move, has long captivated the curiosity of scientists, clinicians, and researchers alike. In this book, *Human Gait – Recent Findings and Research*, we embark on a journey through some of the latest discoveries and advancements in understanding the complexities of human locomotion.

As editors of this compilation, we have sought to bring together a wide variety of perspectives, methodologies, and viewpoints from leading experts in the field. From biomechanics to neuroscience, and from clinical applications to technological innovations, each chapter in this book offers a unique perspective on the multifaceted phenomenon of human gait.

The study of gait crosses disciplinary boundaries, drawing on expertise from fields as diverse as physiology, engineering, rehabilitation, and beyond. As such, this book is a testament to the collaborative nature of scientific inquiry and the collective effort required to unravel the mysteries of human movement.

Whether you are an experienced researcher, a clinician seeking to improve patient care, or simply a curious enthusiast eager to explore the intricacies of the human body, we hope you will find this book a valuable resource. May the insights contained herein inspire further exploration, stimulate new ideas, and ultimately contribute to your collective understanding of one of the most fundamental aspects of human existence: the ability to walk.

We would like to express our sincere gratitude to all the contributors who have generously shared their expertise and knowledge to make this book possible. We hope that *Human Gait – Recent Findings and Research* will serve as a catalyst for continued dialogue, discovery, and innovation in the dynamic field of gait analysis.

Happy reading!

Manuel Domínguez-Morales
Associate Professor at University of Seville,
Architecture and Computer Technology Department,
Seville, Spain

Francisco Luna-Perejón
Postdoc at University of Seville,
Architecture and Computer Technology Department,
Seville, Spain

Section 1

Introduction

Chapter 1

Introductory Chapter: Understanding Human Gait

Manuel Domínguez-Morales and Francisco Luna-Perejón

1. Introduction

Human gait, the intricate biomechanical process by which humans move from one place to another, has been a subject of fascination and study for centuries. From the earliest observations of walking patterns to modern-day biomechanical analysis, understanding human gait has profound implications for fields ranging from sports science to rehabilitation medicine. In this introductory chapter, we delve into recent findings and research surrounding human gait, exploring its complexities and the advancements in our understanding of this fundamental aspect of human locomotion.

2. The basics of human gait

At its core, human gait involves a rhythmic sequence of movements that facilitates locomotion. This sequence typically includes the alternating movements of the lower limbs, coordinated with movements of the upper body to maintain balance and stability. While walking is the most common form of human gait, variations such as running, jogging, and sprinting highlight the adaptability and versatility of this fundamental human function.

3. Biomechanical factors

The study of human gait encompasses a broad range of biomechanical factors, including the mechanics of joints, muscles, and bones involved in locomotion (see **Figure 1**). Several research has elucidated the intricate interplay between these components, revealing the complex mechanisms that govern efficient and effective movement [1]. From the role of muscle activation patterns in propulsion to the biomechanics of foot-ground interaction, advancements in technology and methodology have provided new insights into the biomechanical underpinnings of human gait.

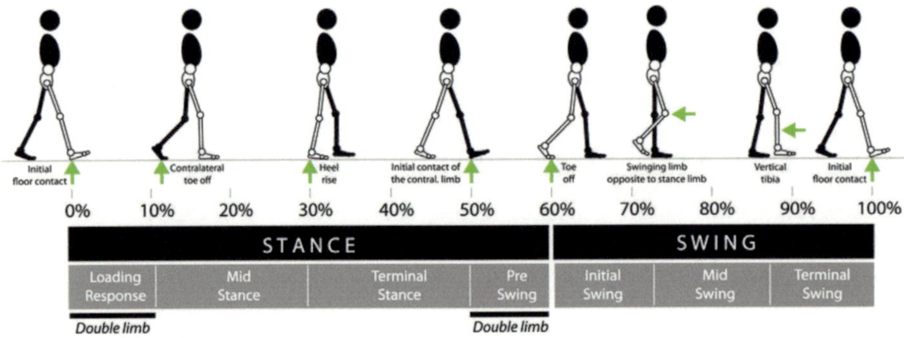

Figure 1.
Illustration showing the biomechanical aspects of human gait.

4. Neural control and coordination

Central to the execution of human gait is the intricate neural control and coordination of movement patterns (see **Figure 2**). Recent research in neuroscience has shed light on the neural pathways and mechanisms responsible for initiating, coordinating, and modulating gait patterns [2]. Understanding the neural basis of gait not only informs our knowledge of normal locomotion but also provides insights into neurological disorders that affect gait, such as Parkinson's disease and cerebral palsy.

5. Clinical applications

The study of human gait extends beyond fundamental research to clinical applications in rehabilitation and healthcare. Recent advancements in gait analysis technology, such as motion capture systems and wearable sensors, have revolutionized the

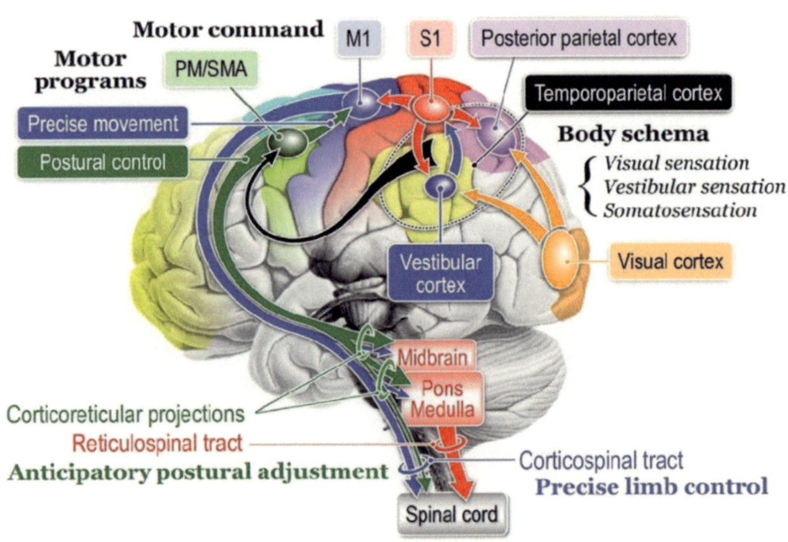

Figure 2.
Brain imaging illustrating neural control of gait.

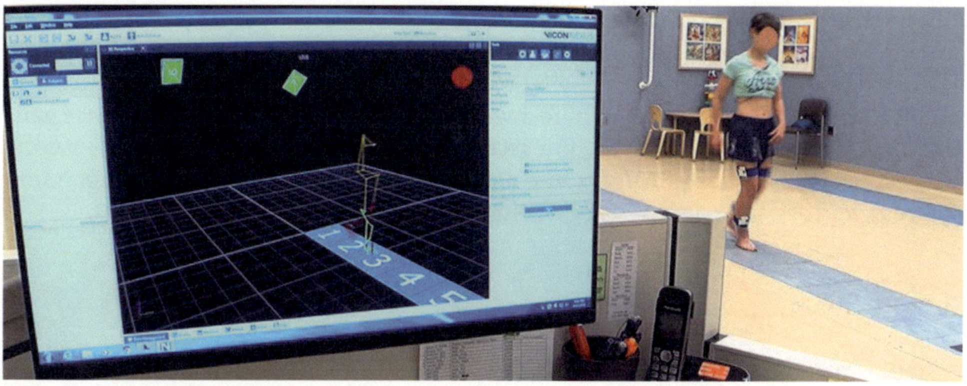

Figure 3.
Motion capture system in a clinical setting.

assessment and treatment of gait disorders [3]. These tools enable clinicians and researchers to quantify gait parameters (see **Figure 3**), identify abnormalities [4, 5], and develop targeted interventions to improve gait function and mobility in patients with neurological and musculoskeletal conditions [6].

6. Future directions

As our understanding of human gait continues to evolve, future research directions hold promise for further advancements in this field. Integrating insights from biomechanics, neuroscience, and clinical practice, researchers are poised to unravel the complexities of human gait and develop innovative approaches to enhance mobility and quality of life for individuals across the lifespan [7].

7. Emerging technologies and methodologies

The exploration of human gait has been greatly enhanced by emerging technologies and methodologies. High-speed motion capture systems, force plates, and computational modeling techniques have provided researchers with unprecedented levels of detail and precision in analyzing gait patterns [8]. Additionally, advances in neuroimaging technologies, such as functional magnetic resonance imaging (fMRI) and electroencephalography (EEG), have enabled researchers to investigate the neural correlates of gait control and coordination in real-time [9].

8. Interdisciplinary collaborations

Collaborative efforts between researchers from diverse disciplines have enriched our understanding of human gait. Interdisciplinary collaborations between biomechanists, neuroscientists, clinicians, and engineers have facilitated the integration of multiple perspectives and methodologies in studying gait [10]. These collaborations have led to innovative research approaches and have accelerated progress in addressing complex questions related to human locomotion.

9. Environmental influences on gait

Environmental factors play a crucial role in shaping human gait patterns. Recent studies have examined how terrain, footwear, and external conditions influence gait dynamics [11]. Understanding the interaction between individuals and their environment can provide valuable insights into designing accessible urban spaces, ergonomic footwear, and assistive devices for individuals with mobility impairments.

10. Conclusion

In this introductory chapter, we have explored the multifaceted nature of human gait, from its biomechanical foundations to its clinical applications and future directions in research. As we delve deeper into the intricacies of human locomotion, we gain a greater appreciation for the remarkable capabilities of the human body and the potential to improve outcomes for individuals with gait impairments. In the chapters that follow, we will delve into specific aspects of human gait research, examining recent findings, methodologies, and implications for theory and practice.

Author details

Manuel Domínguez-Morales* and Francisco Luna-Perejón
Computer Architecture and Technology department, University of Seville, Spain

*Address all correspondence to: mjdominguez@us.es

References

[1] Winter DA. The Biomechanics and Motor Control of Human Gait: Normal, Elderly and Pathological. Vol. 3. Waterloo: Waterloo Biomechanics; 1991

[2] Neptune RR, Kautz SA, Zajac FE. Contributions of the individual ankle plantar flexors to support, forward progression and swing initiation during walking. Journal of Biomechanics. 2009;**42**(6):843-850

[3] Whittle MW. Gait Analysis: An Introduction. London, England: Butterworth-Heinemann Ltd; 1991

[4] Domínguez-Morales M et al. Smart footwear insole for recognition of foot pronation and supination using neural networks. Applied Sciences. 2019;**9**(19):3970

[5] Luna-Perejón F et al. Low-power embedded system for gait classification using neural networks. Journal of Low Power Electronics and Applications. 2020;**10**(2):14

[6] Perry J, Burnfield JM. Gait analysis. Normal and pathological function. 2nd ed. California: Slack; 2010

[7] Zatsiorsky VM, Prilutsky BI. Biomechanics of Skeletal Muscles. Pennsylvania State University Press, Human Kinetics; 2012

[8] Fregly BJ, Besier TF, Lloyd DG, Delp SL, Banks SA, Pandy MG, et al. Grand challenge competition to predict in vivo knee loads. Journal of Orthopaedic Research. 2012;**30**(4):503-513

[9] Clark DJ, Christou EA, Ring SA, Williamson JB. Enhanced somatosensory feedback reduces prefrontal cortical activity during walking in older adults. Journal of Gerontology: Series A. 2014;**69**(11):1422-1428

[10] Graziano MS, Aflalo TN. Mapping behavioral repertoire onto the cortex. Neuron. 2007;**56**(2):239-251

[11] Mickle KJ, Munro BJ, Lord SR, Menz HB, Steele JR. ISB clinical biomechanics award 2009: Toe weakness and deformity increase the risk of falls in older people. Clinical Biomechanics. 2016;**25**(9):767-772

Chapter 2

Perspective Chapter: Hardware Technologies for Gait Restoration

Konstantin V. Lyadov, Elizaveta S. Koneva,
Galina V. Dereviashkina and Vitaly G. Polushkin

Abstract

This chapter summarizes current status and perspectives in hardware technologies for medical rehabilitation. The chapter's first part describes the long journey from basic rehabilitation technologies toward modern robotic devices for gait restoration. The second part of the chapter revolves around a context-based use of hardware techniques: their connection with rehabilitation goals, International Classification of Functioning (ICF) coding, rationale for use, and combinations with other treatment modalities. This part summarizes the opportunities of modern hardware gait and highlights their limitations, both clinical and organizational. The third part revolves around available evidence-based data on the comparative effectiveness of rehabilitation technologies in different clinical scenarios. The final subchapter describes the state-of-the-art hardware restoration techniques, including implanted electrodes, exoskeletons etc., as well as gives an outlook toward the challenges faced with researchers and healthcare professionals.

Keywords: hardware rehabilitation, gait disturbances, poststroke rehabilitation, robotic rehabilitation, weight unloading, biofeedback devices

1. Introduction

Rehabilitation of gait stereotype is one of the most well-established medical rehabilitation interventions with high chance of clinical success. This is due to the large number of rehabilitation technologies and training complexes that have been successfully introduced into clinical practice over the past 25 years [1–4].

Before the first robotic hardware devices for locomotion therapy were introduced in 2001, the restoration of the walking stereotype was carried out by physical rehabilitation professionals using "manual" methods. This was an extremely labor-intensive task and required not only skill but a great deal of stamina. Manual rehabilitation of the patient's walking was usually carried out on bars or on a "treadmill", with the involvement of at least two specialists who provided "manual" assistance in moving the patient's legs. Of course, this technique also had a number of requirements on the patient's side. These requirements included moderate degree of movement disorders, positive postural control, and cognitive and compliance levels sufficient to follow the recommendations during training. The basic means of rehabilitation were four-leg walkers (with high and low support levels), rollators, canes, crutches, single-support canes, the use of which was taught to patients during the course of rehabilitation [5, 6].

This combination of "manual" gait training and the use of technical means of rehabilitation for external support during walking did not allow to effectively restore the physiological stereotype of gait, but rather contributed to general locomotive function.

Modern arsenal of rehabilitation devices and technical means of rehabilitation is extensive. In principle, all technical devices used in practical rehabilitation can be divided into four large groups:

1. systems for unloading body weight;

2. biofeedback systems—reconstruction;

3. robotic walking recovery systems;

4. medical load suits.

Effectiveness and utilization of these technologies per se or in combination with other rehabilitation modalities like electric or magnetic stimulation vary greatly between different clinical contexts, treatment strategies, and rehabilitation prognostic groups; moreover, different aspects of gait can be improved with the use of these techniques [7]. It must be stressed out, that these technologies are not mutually exclusive and can sometimes be combined to treat a single patient. Moreover, sometimes they are intentionally coupled to use their specific traits in combination (e.g., robotic biofeedback systems) [4].

2. Overview of hardware rehabilitation technologies

Body weight unloading systems are devices that allow rehabilitation professionals to securely fix the patient in upright position using a suspension system. Such systems usually include a vest with clamps in the groin area that are tethered to vertical support posts located either above the patient's head or in the side support posts [8–10].

All body weight unloading systems are divided into stationary and mobile models. This division addresses mostly the devise's intended modality of use, not it is technical specification.

The stationary models of the body weight unloading system is a complex technical solution that combines the body weight unloading system and the rehabilitation "treadmill" into a common simulator, with a minimum belt speed of 0.5 km/h, an emergency stop system, often with the ability to read heart rate. These devices are usually installed in medical rehabilitation units. Examples of such devices are "Ortorent C +", "Smart Gravity", "Body weight support system", and "others."

Mobile body weight unloading systems represent a mobile parapodium, with patient fixation and the ability to exercise the patient's walking training directly in the ward or the corridor of the department. As a rule, the appointment of these devices is consistent and at the beginning of the rehabilitation course, the patient begins walking training in the early period in the department, after expanding the motor regimen, the patient is transferred to further training in rehabilitation halls on a stationary simulator with unloading of body weight. These devices are used to tackle two challenges of early rehabilitation—increased risk of falling and need for dosed axial unloading from the lower extremities during walking training, which is often required, especially in patients after musculoskeletal surgery procedures or asthenic patients. Examples include "Ortorent C", "Ortorent M", "Andago", "Aceso", "EGO".

A different ideology of the same technological decision is provided by anti-gravity rehabilitation treadmill "Alter-G", which allows a controlled reduction in the load on the musculoskeletal system, ranging from 20% to 80% of the patient's body weight thanks to differential air pressure technology. During the procedure, the patient enters a special chamber, after which the specialist sets the speed of the walkway, the slope, and air pressure, which gradually increases, gradually removing the axial load on the patient's lower limbs in 1% increments.

In addition, ceiling-mounted "rail" systems for fixing and unloading body weight should also be included in this category of technologies. Often, both patient rooms and rehabilitation rooms are equipped with ceiling "rail" systems. The "rail" system includes a mobile free-sliding "node" to which a patient support device, typically a vest, is attached. Modern rail systems used in the rehabilitation of walking patients are provided with "knots" with a shock-absorbing function that allows physiological training and does not deprive a securely fixed patient of the possibility of physiological training of step elements in conditions of sufficiently rigid vertical fixation. Examples of a "rail" ceiling patient fixation system include GTS (Mega Sarana Medica), Float (Reha Stim), and PRM-01.

BFB-reconstruction systems, in contrast with body weight unloading systems, not only allow for training of the patient's walk skills while maintaining safe fixation and partial body weight unloading but also restore the physiological stereotype of walking through training the elements of the step and walking characteristics in an active motor mode [4]. The most striking example is the C-Mill rehabilitation complex for biofeedback video reconstruction of walking. This rehabilitation complex allows you to play the video sequence on the canvas of the track, individualizing the training and taking into account the pathological signs of the patient's walking. Video marks allow you to adjust such characteristics of the pattern as the length and width of the step, to train complexly coordinated walking elements. Video-audio training support allows physicians to set a goal for rhythmic steps with the same length and width of the step of the contralateral limbs. In addition, this rehabilitation system by means of an integrated podometric diagnostics system, allows the specialist to carry out not an empirical selection of a technique for training the walking stereotype but to conduct initial and dynamic testing of the patient to determine the key gait abnormality features. This examination can be followed by individualization of the walking recovery technique based on the results of the diagnostic study. The rehabilitation complex is also provided with a system for fixing and unloading the patient and means of emergency stops.

Both body weight unloading and BFB-reconstruction systems, however, rely greatly on patient's cognitive skills and his/her ability to cooperate with rehabilitation team. They also require some initial patient's training, and creating user-friendly biofeedback mechanisms is considered a top research priority [2].

Robotic systems for restoring the stereotype of walking. The era of automated imposition of a physiological walking pattern by mechanotherapy in robotic rehabilitation complexes was laid by the Swiss company Hocoma, a medical engineering company that develops innovative equipment for rehabilitation, the first robotic product of which was the Lokomat System [3]. This fundamentally new rehabilitation device appeared on the market in 2001 and opened the era of robotic rehabilitation devices, which has become not only a new direction in the development of high-tech rehabilitation care but also made it possible to train walking in patients with profound motor disorders, including patients with lower paraplegia, hemiplegia, and tetraplegia, which was impossible before the advent of this equipment. Thus, the creation of the Lokomat System, after many years of research and development, has brought a significant breakthrough in the field of locomotor therapy. The Lokomat System is a hardware complex consisting of three parts: external robotic exoskeletons, structurally imitating the lower limbs (motorized nodes reproduce the hip and knee joints, and the "stirrup" fixes the patient's foot). The technical solution of the exoskeleton allows not only to provide rigid fixation of the legs of patients but also to set the necessary anthropometric, goniometric, and speed characteristics, thereby individualizing the training procedures, taking into account the anatomical and physiological characteristics of each patient. The patient unloading system provides maximum technical safety for patient verticalization, as well as the possibility of providing dosed axial unloading. The constructive solution of the treadmill path, along which the patient walks, allows you to start training at a minimum speed (1 km/h). The operation of the exoskeleton is synchronized with the speed of the track. The software allows not only walking training in the passive mode but also dosed to reduce the degree of functional activity of the robot (alternately or simultaneously) from the patient's legs, thereby conducting training in the active motor mode, subject to the imposition of a physiological walking stereotype on the patient. Currently, in clinical practice around the world, there is a sufficient number of analog robotic systems: robotic complexes "GEO" and "A3". The principle of operation of all robotic rehabilitation devices is based on the imposition of a physiological stereotype of walking by exoortheses in the patient's fixation system. However, nevertheless, there are a number of differences, so the GEO complex, unlike the Lokomat System, allows you to train walking not only in a straight line but also to simulate going up and down stairs, thereby expanding the possibilities of using the technology in patients of different nosological forms and patterns of gait disturbance.

The next type of robotic technology that has also been developed in the Hocoma industrial laboratory is the Erigo robotic stander. The turntable with an integrated stepping device has been designed to facilitate and facilitate the early mobilization of neurological patients. This hardware complex will allow for staged verticalization of patients with a given verticalization angle, and an integrated robotic stepping device that starts mobilization of the patient's lower extremities in a horizontal position allows for more effective verticalization with a reduced risk of collaptoid conditions. The device securely fixes the patient in a corset in the body weight unloading system on the horizontal surface of the verticalizing table with setting the verticalization angle; the integrated robotic stepping device fixes the distal parts of the lower extremities and allows the specialist to set the walking speed and rhythm parameters. The software creates and saves individual reports of verticalization procedures, allowing for dynamic monitoring of the effectiveness of the ongoing course of procedures with a logged report form. An analog example of this type of equipment is the hardware complex "A1".

Exoskeleton (from the Greek. ε'ξω—external and σκελετος—skeleton) is a device designed to increase human strength by means of external frame. Initially developed for the military, they now reliably entered the rehabilitation and habilitation practice. There are models of exoskeletons with an active and passive principle of operation—active and passive exoskeletons. Active models use external devices as a source of energy, while the mechanics of passive exoskeletons are based on the use of kinetic energy and human strength. Active exoskeletons are widely used for military purposes, but their performance data is rarely open to public or non-military researchers. The maximum number of such developments falls on the Pentagon. One of the well-known exoskeletons HULC (Lockheed Martin, USA), allows a soldier to quickly move with a load over rough terrain, while there is a high speed of movement. HULC helps not only carry but also lift the load from the ground. The mass of the device is 25 kg, most of it falls on batteries, and the battery lasts for two hours. From a medical perspective, their number also increases; medical exoskeletons can be divided into two groups.

The first group completely the ReWalk exoskeleton (ARGO Medical Technologies, Israel) allows patients with lower paraparesis to stand up and walk using sticks. The operation of the structure is based on sensors that detect the forward tilt of the body and transmit a signal to the devices supporting the legs. Power is supplied from a battery placed in a special backpack behind your back. The design can only be used in persons with preserved functions of the upper limbs. REX (REX Bionics, New Zealand)—provides additional support for the human body when moving. Management is carried out using a joystick and a tablet. The mass of the exoskeleton is 38 kg, which, together with the high cost, greatly limits its wide-scale use. HAL—Hybrid Assistive Limb (Cyberdyne, Japan)—is designed for the elderly and disabled people who have difficulty in moving. The total weight of the structure is 23 kg, height is 160 cm. In addition, the battery weighs 10 kg, and the battery life (under maximum load conditions) is 2.5 hours. These products mostly focus on patients that have little to no perspective of returning to normal gait stereotype either due to excessive skeletal trauma or as a result of spinal cord disruption or specific brain trauma.

eLEGS (Ekso Bionics, USA) is a special hydraulic exoskeleton designed for patients with lower paraparesis. The design allows them to move around with the help of crutches or special walkers. The heart of the machine is an interface-hardware-software complex that uses natural human movement to safely translate it into exoskeleton action using a microcomputer. Passive exoskeletons have found their primary use in military applications. In Russia, the Transport Walking Systems company has created a passive exoskeleton K-2, designed for the needs of the Ministry of Emergency Situations. This device helps a person to carry loads weighing up to 50 kg for a long time without much effort and load on their own musculoskeletal system. A group of Russian scientists from the Research Institute of Mechanics, Moscow State University. M.V. Lomonosov, a working sample of the ExoAtlet P passive modification exoskeleton for rescuers of the Ministry of Emergency Situations has been created, which allows the human operator to carry large loads (70–100 kg). The use of exoskeletons in practical rehabilitation is quite widespread; however, it is still limited due to the severity of the device and the difficulty of walking in it, taking into account the need to maintain balance while relying on crutches. Thus, exoskeletons should be considered to a greater extent in the aspect of assistive technologies for spinal patients as an opportunity for their mobilization and habilitation. At the same time, stationary robotic complexes such as Lokomat are more promising for training a dynamic walking pattern. At the same time, it should be noted that in pediatric practice, on the contrary, the use of the ExoAtlet Bambini exoskeleton is seen as extremely promising for the rehabilitation of children and adolescents with locomotion disorders. Training in a given physiological pattern of walking with the help of a specialist, without additional support on crutches, in combination with the lightweight design of the exoskeleton itself, makes it possible to conduct locomotor training of children in an automated imposition of a stereotype.

The systems of robotic imposition of the walking stereotype also include simulators with ellipsoid step automation. These robotic step simulation systems differ from the previously described complexes by their smaller overall dimensions, lower cost, and ease of operation. The simulators represent an automated simulator of the "ellipse" type, with a system of dynamic unloading of body weight and software that allows you to implement the stereotypical act of automating the patient's walking forward and backward, at a given walking speed and taking into account the patient's anthropometric parameters. Training complexes have a built-in control of biological indicators, control of the angle of elevation of the step, step length, as well as resistance during movement. An example of a training complex: "Ortorent C++".

As clinical, academic, or pragmatic guidelines do not exist for use of exoskeletons, there is much discussion on their role and effectiveness in clinical practice. Recent studies suggest exoskeletons can benefit inpatients with gait impairment [11].

The fourth type of rehabilitation equipment is therapeutic power suits, which is a powered system consisting of supporting elements and adjustable rods, with the help of which, for a therapeutic purpose, a load is imposed on or lifted from the musculoskeletal system [12]. This type of rehabilitation device was created on the basis of developments in space medicine and, in particular, the Penguin suit, designed to protect astronauts from the adverse effects of weightlessness. Therapeutic power suits are both a soft orthopedic apparatus and a load simulator. As an orthopedic device, it contributes to the simultaneous correction of the posture, bringing the joints to the maximum physiological position. As a load suit, it helps to extinguish pathological reflexes and dose the load, enhancing the effect of therapeutic exercises. An extremely important therapeutic effect from the use of costumes is the elimination of the pathological tone that prevents the implementation of the physiological locomotor act. These medical suits are widely used in neurorehabilitation, especially in pediatrics. An example of this technology is the Adele medical suit. Another example of a medical suit is the reflex-loading device "Gravistat". Numerous rods set an adjustable compression load directed along the long axis of the body and correct the position of individual motor segments of the trunk and lower extremities. Overcoming additional resistance increases the activity of postural muscle groups. Axial load and functional correction of the position of motor segments of the body by rotational rods leads to the emergence of impulses from the receptors of the articular-muscular-ligamentous apparatus to the central nervous system. The third representative of the family of medical suits is the neuro-orthopedic rehabilitation pneumosuit "Atlant", developed on the basis of the high-altitude compensating flight suit VKK-6, which was part of the equipment of pilots and astronauts. RPK "Atlant" promotes polysegmental stretching of the muscular-ligamentous apparatus and activation of the motor-neuronal system at all levels of the central nervous system. The Atlant suit creates neurophysiological conditions that allow the patient to maintain a posture, perform voluntary and coordinated movements.

3. Clinical use context

An important aspect of the rehabilitation of walking is the definition of its place in the structure of the International Classification of Functioning (ICF), for the correct positioning of the role of mechanotherapeutic restoration of the locomotor act in complex programs for the rehabilitation of patients. Gait disorders are specified as the third detailing level in ICF and are defined by four main domains:

- d4500: walking short distances. According to the WHO comments on domains—walking for distances less than a kilometer, for example, in a room, corridor, short distances outside the home;

- d4501: long distance walking. According to the WHO comments on domains—walking distances of more than a kilometer, for example, from one part of a village or city to another;

- d4502: walking on various surfaces. According to WHO comments on domains—walking on sloping, uneven, moving surfaces such as grass, ice, gravel, snow, walking on the deck of a ship, train, or other vehicle;

- d4503: walking around obstacles. According to WHO comments on domains—walking between moving and stationary objects, among people, animals and vehicles, walking in the market shop, in traffic conditions or in crowded places.

An important element of gait rehabilitation in order to individualize training methods is the diagnosis and analysis of the pattern, both at the beginning of the course and in the dynamics of training [13]. For example, the use of the diagnostic tool of the previously described C-Mill biofeedback videoreconstruction complex allows specialists to carry out the necessary diagnostics without additional equipment and with relaxed skill requirements. However, pre-installed diagnostic tools are found only in selected hardware complexes, so it is usually necessary to take into account the need for diagnostics using the following technical means and conditions:

- a sufficiently large free area to assess the free gait of the patient (at least 6 meters in length and 3 meters in width);

- "running" track;

- stopwatch for recording training time and fixing heart rate;

- a video camera and a computer for "cutting" video fragments;

- software for 2D analysis;

- plantar pressure recording system;

- power platform and vector video display unit;

- four- or eight-channel surface myographic amplifier;

- 3D motion analysis system;

- breath analyzer.

Rehabilitation goals for walking restoration programs should be determined by the domains of the ICF. Increasing the distance and speed of walking, as well as, and these are priority goals for restoring the functioning of the patient and improving his quality of life—this is the development of complexly coordinated elements of walking, and overcoming obstacles and confident walking on different surfaces, including unstable ones. All of the above tasks should be realized under the condition of restoring the elements of the step and restoring the global physiological stereotype of walking. These goals, when compiling complex rehabilitation programs, are formed sequentially: first, the distance and speed of walking are trained, then the tasks of adapting the patient to the social and domestic environment and training walking with obstacles on various surfaces, in combination with the implementation of complex functioning skills, are already solved, including the lack of visual control over the implementation of walking. In this regard, an important component in the staged implementation of walking recovery is the inclusion of ergotherapeutic walking training technologies in the rehabilitation program, namely the restoration of walking, as a component of the implementation of a certain social, household, or labor skill. An example of such a complex rehabilitation with recovery according to the domains of the ICF on the functions of overcoming obstacles and walking on various surfaces was created by the author's team. In this training device, the BFB principle is implemented by distracting the patient's focus on steps with audio and visual. In this rehabilitation complex, the multidimensional reproduction of social and domestic tasks is carried out by modeling various situations either on the screen in front of the patient or using VR technologies with the patient's maximum immersion in the "real" environment. At the same time, both ergotherapy items (locks, door handles, magnetic card readers, etc.) and unstable surfaces that make it difficult to walk on "reference" rehabilitation surfaces are implemented in the complex training system [14].

When discussing the means of walking rehabilitation, one cannot fail to note the importance of a staged and integrated approach to the issue of restoring this complex locomotor act. The phasing of the use of mechanotherapeutic means of rehabilitation is reflected in the successive change of rehabilitation devices, taking into account the degree of verticalization and the possibility of reproducing motor acts in an active motor mode. As a rule, the following sequential system of using the described equipment occurs. Hardware verticalization, followed by the appointment of either robotic systems (in case of deep motor deficits) or biofeedback video construction in active motor mode (in case of a shallow motor deficit), with a gradual transition to body weight unloading systems for walking training without the condition of imposing a physiological pattern. This staged approach is more typical for patients with neurological deficits, while patients with lesions of the musculoskeletal system, as a rule, are limited to the appointment of devices with unloading of body weight and/or biofeedback reconstruction and undergo rehabilitation of walking in the active motor mode, subject to axial unloading of the lower extremities.

In clinical practice, the use isolated methods of mechanotherapy are not feasible. The restoration of the walking stereotype has a systematic and integrated approach [7, 15, 16]. In complex walking rehabilitation programs, in addition to individual kinesiotherapy sessions to work out individual elements of the step and strengthen certain muscle groups, balance training using stabiloplatforms, including BFB for training balance and locomotor symmetry, is prescribed, for example, "Huber", MBN "Stabilo", and others. A simulator that simulates climbing stairs with an electric drive is widely prescribed for training getting up and standing, for training walking up and down stairs, and for the ability to train walking on steps of different heights. An example of this equipment can be: "DST 8000 Triple Pro", "Alterstep". In the vast majority of cases, mechanotherapeutic rehabilitation of the walking stereotype is carried out in combination with electromyostimulation technologies. Special attention should be paid to "Walk Aid", a programmable wearable electrical stimulator, widely used in foot paresis, which gives an impulse to the "transfer" phase in the step sequence, thereby allowing patients with peripheral paresis to carry out the physiological walking pattern. The technologies of multichannel functional electrical stimulation are also widely applicable, which make it possible to enhance rhythmic muscle activation during walking training. Examples of devices—"Akord", "Trust-M", "MNS-16", and others.

4. Current evidence on best training system utilization

Many fucking devices—lots of techniques to choose. Aggregated data suggests that both the functional deficit's properties and the clinical diagnosis contribute to the desired properties of hardware rehabilitation technique. Studies vary greatly in quality of evidence and patient groups. Thus, to summarize briefly the overall benefits of hardware rehabilitation devices, we have extracted data from meta-analysis of related studies and specifically discussed the issues of study design quality, study data applicability, and limitations. The systematic reviews are listed in **Table 1**.

It is important to stress the key limitations of these studies. They can be divided into three overlapping groups: related to the nature of the underlying medical condition, related to the structure of patient management, clinical logistics, and medical aid timing, and related to the technical characteristics of the devices in question and their utilization context.

In the first group, the associated results can significantly vary with time, which is especially important in cases with specific disabling events like stroke or trauma; majority of studies describe late subacute (3–6 months) or chronic (>6 months) period after stroke or trauma; studies focusing on acute and early subacute patients

№	Author and date	Patient group	Compared modalities	Conclusion
1	Moucheboeuf 2020 [3]	Post-stroke patients	Robotic vs. physiotherapy vs. physiotherapy combined with body weight support training	Robot-assisted training resulted in greater improvement of gait speed and balance
2	Yang, 2022 [17]	Spinal cord injury (N = 497)	body weight-supported overground training vs. body weight-supported treadmill training vs. robot-assisted gait training vs. conventional training	Each hardware modality is better than conventional physiotherapy, but no studies compare them with each other
3	Glanz, 1995 [18]	Poststroke (chronic) patients with hemiplegia	Biofeedback plus physiotherapy vs. physiotherapy alone	No increase in range of motion of speed
4	Hsu, 2023 [19]	Post-stroke, any time after the event (N = 492)	Exoskeleton-assisted training vs. conventional gait training	Exoskeleton-assisted training is superior in walking speed and balance
5	Mehrholz, 2017 [20]	Post-stroke, any time after the event (N = 1472)	Electromechanical- and robotic-assisted gait training versus physiotherapy	Hardware-assisted training was associated with increased chances to return to independent walk
6	Patathong, 2023 [1]	Incomplete spinal cord injury (N = 709)	Physical therapy vs. treadmill vs. functional electrical stimulation vs. robotic-assisted gait training,	Functional electrical stimulation is the best in regaining speed and velocity; no large studies available

Table 1.
Systematic reviews and meta-analysis on hardware rehabilitation technologies.

are rare [2]. For example, of all biofeedback studies, only Druzbicki et al. used step-length biofeedback during bodyweight-supported treadmill walking [21].

Second, the studies' results may be associated with specific patterns of available equipment in different hospitals. The most impressive example is the meta-analysis by Yang et al. [17], where 15 studies compared some hardware rehabilitation technology with physiotherapy, but no comparative studies between hardware modalities were found. It is also worth mentioning that specific hardware modalities themselves are quite multifarious by regimen, intensity of use and specific operating modes, making comparison between this embedded treatment options not less importan than comparison between different modalities.

Last, but not the least, technical properties of hardware rehabilitation devices and selected modalities also affect the treatment outcomes. Rehabilitation programs include not only the hardware devices themselves but also the associated gait restoration techniques, and these variations in associated training can contribute to the hardware rehabilitation results.

In the end, available comparative data on hardware gait rehabilitation techniques, especially in high-quality RCTs, is very limited. The enormous variability of devices and their use options, multiplied by the fact that hardware rehabilitation cannot be the single rehabilitation intervention in these patients, as well as the heterogeneity of patient groups, greatly restrict the quality of evidence in this field.

Studies may suggest that careful selection of patients with poststroke gait disorders for specific modalities of exoskeleton and robotic rehabilitation can improve walking speed and balance outcomes compared to modalities. In patients with spinal cord injury bodyweight-supported overground training, bodyweight-supported treadmill training and robot-assisted gait training are all more effective than conventional training, but no comparative research between them exists [17].

More specific assumptions arise not from meta-analysis but from individual studies. A large article by Mikolajczyk et al. recaps the available data on specific gait rehabilitation modalities in stroke, spinal cord injury, Parkinson's disease, and multiple sclerosis with respect to speed, stability, and independence of walk aspects [22].

These details suggest that the following use context. In any individual patient, the gait disorder pattern should be decomposed to components specific to known benefits of available hardware rehabilitation devices (e.g., walking speed, balance, or independent walking ability). This can help the clinicians to prioritize one technology over the other for this particular patient while staying inside an evidence-based context.

In our opinion, more real-world data would suggest a more pragmatic and, in the meantime, broad look at hardware gait rehabilitation. Internet of things and artificial intelligence analysis will finally reduce the workload demands for new research, providing rehabilitation society with new quality evidence on this topic.

5. Technological perspectives

First in line of emerging technologies come exoskeletons, who is clinical role, as described above, is still a subject to discussion with a gradual increase of general clinician acceptance [11]. Their current use is limited by high costs and high weight to inpatient settings only. Upscaling of production, use of new materials, and increased battery useful life can also broaden their practical spectrum.

Minituarization offers another benefit for gait restoration like outpatient-based gait analysis devices. While they are focused mainly on measuring Parkinson's disease gait response to treatment, its possible applications in rehabilitation, especially home-based rehabilitation (see below), could not be underestimated [7]. In the meantime, biofeedback devices earn increasing popularity for poststroke gait rehabilitation as systematic reviews give more positive results on their effectiveness [2].

A short-term technological perspective of hardware rehabilitation encompasses a broad spectrum of AR and VR technologies. As a part of the size-reducing initiative, augmented reality technologies can greatly enhance both inpatient and outpatient performance of gait rehabilitation [23]. However, literature reviews often focus on academic, rather than technological perspectives of these technologies [24].

Deep-tech perspectives of gait restoration lie mostly in the field of neural stimulation and neural implanting. Generally, both spine and peripheral installments show promising clinical results, and some randomized controlled trials are already available – for example, on peroneal nerve stimulation [25]. However, more interesting results are shown in studies, yet single patients describe brain-spine implantable devices [26]. Astonishing case reports with one complete walk restoration after complete spinal injury indicate that this stimulation devices can possibly lead to real neural regeneration – at least turning the device after a prolonged period of use did not compromise the restored functions [27].

However, even wider use of such technologies does not reduce the need for more "classical" hardware rehabilitation technologies. First, they are currently limited to use in patients with relatively low spinal injuries and preserved cognitive function. Second, the rehabilitation process still require a "training" phase, where the patient learns to control his "newly-acquired" locomotion skills. Last, but not the least, patients with leg trauma constitute a substantial proportion of all patients requiring gait restoration treatment.

Conflict of interest

The authors declare no conflict of interest.

Author details

Konstantin V. Lyadov, Elizaveta S. Koneva, Galina V. Dereviashkina and Vitaly G. Polushkin*
Sechenov University, Moscow, Russia

*Address all correspondence to: polushkinvp@gmail.com

References

[1] Patathong T, Klaewkasikum K, Woratanarat P, Rattanasiri S, Anothaisintawee T, Woratanarat T, et al. The efficacy of gait rehabilitations for the treatment of incomplete spinal cord injury: A systematic review and network meta-analysis. Journal of Orthopaedic Surgery and Research. 2023;**18**(1):60

[2] Spencer J, Wolf SL, Kesar TM. Biofeedback for post-stroke gait retraining: A review of current evidence and future research directions in the context of emerging technologies. Frontiers in Neurology. 2021;**12**:637199

[3] Moucheboeuf G, Griffier R, Gasq D, Glize B, Bouyer L, Dehail P, et al. Effects of robotic gait training after stroke: A meta-analysis. Annals of Physical and Rehabilitation Medicine. 2020;**63**(6):518-534

[4] Pinheiro C, Figueiredo J, Cerqueira J, Santos CP. Robotic biofeedback for post-stroke gait rehabilitation: A scoping review. Sensors (Basel). 2022;**22**(19). Available from: https://www.ncbi.nlm.nih.gov/pmc/articles/PMC9573595/

[5] Fricke SS, Bayón C, der Kooij HV, FvA EH. Automatic versus manual tuning of robot-assisted gait training in people with neurological disorders. Journal of Neuroengineering and Rehabilitation. 2020;**17**(1):9

[6] Conesa L, Costa Ú, Morales E, Edwards DJ, Cortes M, León D, et al. An observational report of intensive robotic and manual gait training in sub-acute stroke. Journal of Neuroengineering and Rehabilitation. 2012;**9**:13

[7] Salchow-Hömmen C, Skrobot M, Jochner MCE, Schauer T, Kühn AA, Wenger N. Review-emerging portable technologies for gait analysis in neurological disorders. Frontiers in Human Neuroscience. 2022;**16**:768575

[8] Kim K, Chong WS, Yu CH. Rehabilitative training system based on a ceiling rail for detecting the intended movement direction of a user. Technology and Health Care. 2020;**28**(S1):443-452

[9] Shibata Y, Imai S, Nobutomo T, Miyoshi T, Yamamoto S. Development of body weight support gait training system using antagonistic bi-articular muscle model. Annual International Conference of the IEEE Engineering in Medicine and Biology Society. 2010;**2010**:4468-4471

[10] Frey M, Colombo G, Vaglio M, Bucher R, Jörg M, Riener R. A novel mechatronic body weight support system. IEEE Transactions on Neural Systems and Rehabilitation Engineering. 2006;**14**(3):311-321

[11] Gillespie J, Arnold D, Trammell M, Bennett M, Ochoa C, Driver S, et al. Utilization of overground exoskeleton gait training during inpatient rehabilitation: A descriptive analysis. Journal of Neuroengineering and Rehabilitation. 2023;**20**(1):102

[12] Louie DR, Eng JJ. Powered robotic exoskeletons in post-stroke rehabilitation of gait: A scoping review. Journal of Neuroengineering and Rehabilitation. 2016;**13**(1):53

[13] Nonnekes J, Nieuwboer A. Towards personalized rehabilitation for gait impairments in Parkinson's disease. Journal of Parkinson's Disease. 2018;**8**(s1):S101-S1s6

[14] Derevyashkina GV, Koneva ES, Shapovalenko TV, Bisheva DR, Sidyakina IV, Konev SM, et al. Recovery of social and everyday skills after a complex of functional spatially oriented

rehabilitation in elderly patients with cerebral stroke. Voprosy Kurortologii, Fizioterapii, i Lechebnoĭ Fizicheskoĭ Kultury. 2022;**99**(4. Vyp. 2):5-10

[15] Skvortsova VI, Ivanova GE, Rumiantseva NA, Staritsyn AN, Kovrazhkina EA, Suvorov A. Modern approach to gait restoration in patients in the acute period of cerebral stroke. Zhurnal Nevrologii i Psikhiatrii Imeni S.S. Korsakova. 2010;**110**(4):25-30

[16] Dickstein R. Rehabilitation of gait speed after stroke: A critical review of intervention approaches. Neurorehabilitation and Neural Repair. 2008;**22**(6):649-660

[17] Yang FA, Chen SC, Chiu JF, Shih YC, Liou TH, Escorpizo R, et al. Body weight-supported gait training for patients with spinal cord injury: A network meta-analysis of randomised controlled trials. Scientific Reports. 2022;**12**(1):19262

[18] Glanz M, Klawansky S, Stason W, Berkey C, Shah N, Phan H, et al. Biofeedback therapy in poststroke rehabilitation: A meta-analysis of the randomized controlled trials. Archives of Physical Medicine and Rehabilitation. 1995;**76**(6):508-515

[19] Hsu T-H, Tsai C-L, Chi J-Y, Hsu C-Y, Lin Y-N. Effect of wearable exoskeleton on post-stroke gait: A systematic review and meta-analysis. Annals of Physical and Rehabilitation Medicine. 2023;**66**(1):101674

[20] Mehrholz J, Thomas S, Werner C, Kugler J, Pohl M, Elsner B. Electromechanical-assisted training for walking after stroke. Cochrane Database of Systematic Reviews. 2017;**5**(5):Cd006185

[21] Drużbicki M, Przysada G, Guzik A, Brzozowska-Magoń A, Kołodziej K, Wolan-Nieroda A, et al. The efficacy of gait training using a body weight support treadmill and visual biofeedback in patients with subacute stroke: A randomized controlled trial. BioMed Research International. 2018;**2018**:3812602

[22] Mikolajczyk T, Ciobanu I, Badea DI, Iliescu A, Pizzamiglio S, Schauer T, et al. Advanced technology for gait rehabilitation: An overview. Advances in Mechanical Engineering. 2018;**10**(7):1687814018783627

[23] Khokale R, Mathew GS, Ahmed S, Maheen S, Fawad M, Bandaru P, et al. Virtual and augmented reality in post-stroke rehabilitation: A narrative review. Cureus. 2023;**15**(4):e37559

[24] Kiani S, Rezaei I, Abasi S, Zakerabasali S, Yazdani A. Technical aspects of virtual augmented reality-based rehabilitation systems for musculoskeletal disorders of the lower limbs: A systematic review. BMC Musculoskeletal Disorders. 2023;**24**(1):4

[25] Hutin E, Ghédira M, Vinti M, Tazi S, Gracies JM, Decq P. Comparing the effect of implanted peroneal nerve stimulation and ankle-foot orthosis on gait kinematics in chronic hemiparesis: A randomized controlled trial. Journal of Rehabilitation Medicine. 2023;**55**:jrm7130

[26] Makowski NS, Lombardo LM, Foglyano KM, Kobetic R, Pinault G, Selkirk SM, et al. Walking after incomplete spinal cord injury with an implanted neuromuscular electrical stimulation system and a hinged knee replacement: A single-subject study. Spinal Cord Series And Cases. 2020;**6**(1):86

[27] Lorach H, Galvez A, Spagnolo V, Martel F, Karakas S, Intering N, et al. Walking naturally after spinal cord injury using a brain-spine interface. Nature. 2023;**618**(7963):126-133

Section 2

Applications

Chapter 3

Automated Camera-Based Assessment of Short Physical Performance Battery (SPPB) for Older Adults with Cancer

Larry Duncan, Shaotong Zhu, Mackenzi Pergolotti, Smith Giri, Hoda Salsabili, Miad Faezipour, Sarah Ostadabbas and S. Abdollah Mirbozorgi

Abstract

This chapter introduces a motorized camera system designed for monitoring and evaluating the tests of the Short Physical Performance Battery (SPPB). This system targets physical performance assessment for older patients undergoing cancer treatment. The device is self-contained, incorporating a small computer, three cameras, and two motors. The core algorithms utilize three object recognition techniques: template matching, Haar cascades, and Channel and Spatial Reliability Tracking. To facilitate user interaction, graphical user interfaces (GUIs) are developed on the small computer, enabling test execution and camera adjustments via cell phone and its hotspot. The system delivers precise results, with gait speed tests showing a range of 0.041–1.92 m/s and average speed and distance accuracies exceeding 95%. The standing balance and 5 times sit-stand (5TSS) tests achieve average time accuracies exceeding 97%. This novel camera-based device holds promise in enhancing evaluation of lower body extremity fitness for elderly people receiving cancer health care, offering a reliable and efficient solution for monitoring their progress and well-being.

Keywords: 5 times sit-stand, gait speed, image processing, object recognition, Short Physical Performance Battery (SPPB), standing balance

1. Introduction

Approximately 60% of new cancer diagnoses and 70% of cancer-related deaths are found in adults aged 65 years and above [1]. The number of cancer survivors in the United States is expected to exceed 20 million by 2026 [2], with many experiencing a negative impact on their physical, social, and cognitive quality of life, potentially leading to disability [3]. Older adults undergoing cancer treatment are particularly vulnerable to functional decline, disability, increased healthcare use, and increased time in the hospital [4]. To address these challenges, there is a pressing need to fine-

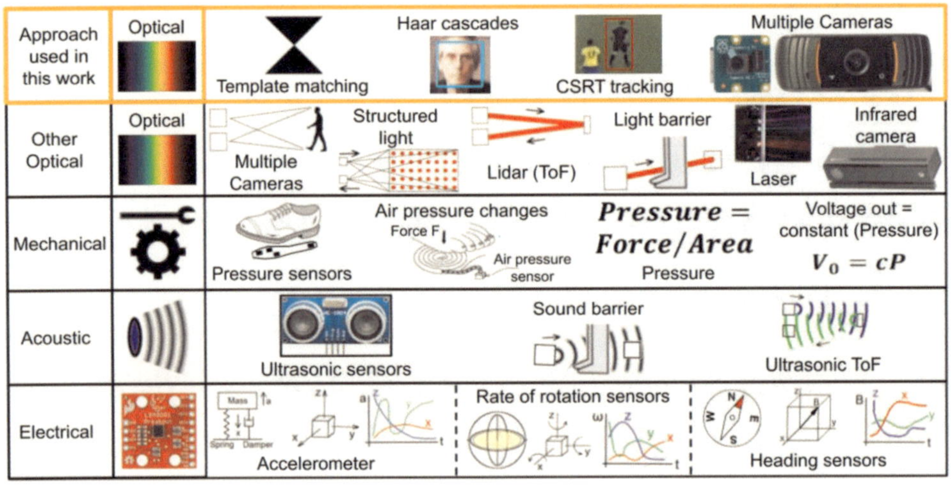

Figure 1.
An overview of various technologies employed for gait speed, standing balance, and 5 times sit-stand tests. Row 1 presents the camera design implemented in this study, utilizing three object recognition techniques. The small square camera is the Raspberry Pi Camera V2.1 [11]. Row 2 shows camera and optical sensing systems, featuring lasers [12] and an infrared camera (Xbox One Kinect) [13]. Row 3 demonstrates mechanical systems, including shoe and air bladder sensors [14]. Row 4 includes ultrasonic sensor systems [15]. Row 5 has electrical accelerometer, gyroscope, and magnetometer sensors [16].

tune cancer treatment for older adults by evaluating their physical fitness and estimating treatment tolerability [5].

Tests of fitness in physical activity, such as the Short Physical Performance Battery (SPPB), are recommended for determining treatment tolerability, but they are not routinely included in normal clinical practice due to limitations of time and manual testing [6, 7]. The SPPB, which objectively measures lower extremity function, includes tests for walking speed, time to perform 5 chair-stands, and 10-second standing balance tests [8].

To overcome the limitations of manual SPPB assessments and improve starting evaluation, treatment preparation, and progress monitoring, automated tools are needed [9, 10]. In this context illustrated by **Figure 1**, vision-based systems, force-based mechanical approaches, acoustic sensor systems, and microelectromechanical systems (MEMS) have been explored. Especially with vision-based systems, much data can be collected [17–19].

In this work we propose a self-standing Camera-based Body Motion Tracking (CBMT) system, inspired by earlier studies [20, 21], and designed for use in clinics and hospitals to enable easier fitness evaluation of patients using the SPPB tests. The CBMT system performs object recognition and tracking of people, accurately processes all SPPB statistics, can be monitored via a cell phone, and allows for report generation and cloud storage for medical professionals' access. The design overview, hardware and software details, software implementation, experimental results, and a comparison with other systems are presented in subsequent sections. The chapter concludes with a discussion and implications of the proposed design.

2. Hardware design and overview

We developed a portable and standalone platform using a Raspberry Pi (RPi) computer instead of a regular computer for its compact size and ability to control multiple cameras and motors without adding other devices. **Figure 2** illustrates the process of running tests with the CBMT system, depicting the coverage angles and range for the gait speed, standing balance, and 5 times sit-stand (5TSS) tests of the SPPB. This platform enables the evaluation of physical performance and supports personalized cancer treatment for older adults.

The CBMT's cameras have an optimal detection range of 2–7.2 m, with a 60° viewing angle and a total horizontal rotation of 180°, providing a total view angle of 240°. The motors can also tilt the cameras up and down if required (±20°).

Figure 3 presents the block diagram representing the connecting operations of the hardware and software of the proposed CBMT system. It consists of two main parts: 1) the camera system (comprising the Raspberry Pi computer, motors, and cameras) and 2) the control smartphone that remotely operates the RPi and the system.

The CBMT platform's hardware components include: 1) one RPi 3B+ computer with 1 GB of RAM, 2) two DC motors, 3) two driver boards (L9110H) for the motors, 4) two USB cameras (Zealinno 1080P Webcam), 5) one Raspberry Pi V2.1 camera module, 6) a few acrylic sheets for assembly, 7) two camera fine adjustment pieces, and 8) a tripod. This physical setup performs all the necessary functionalities, with a

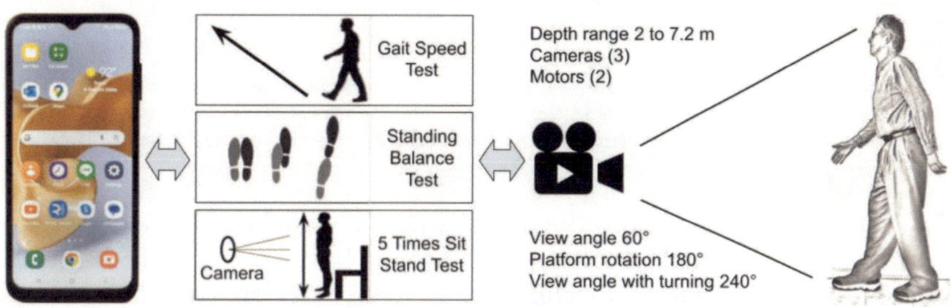

Figure 2.
Process of camera-based device performing short physical performance battery tests.

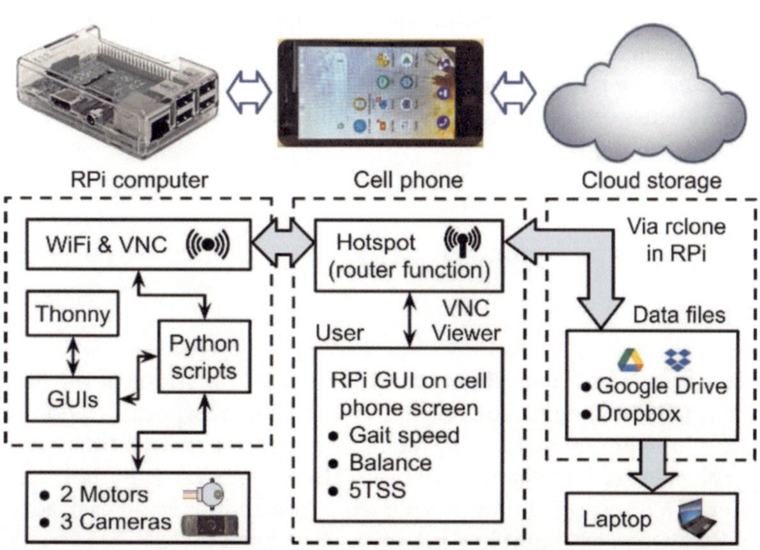

Figure 3.
The block diagram outlining the interconnections and interactions between the hardware and software components of the CBMT system.

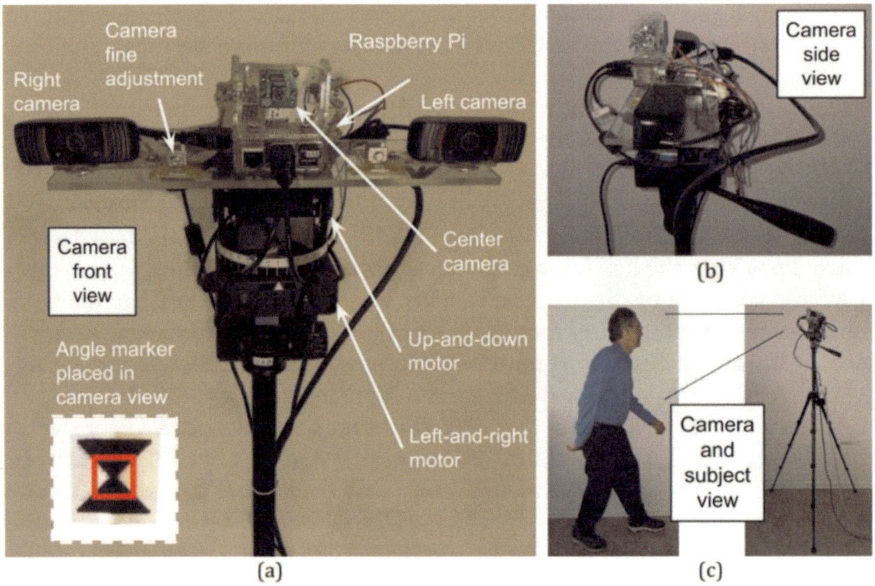

Figure 4.
The fully implemented platform features a Raspberry Pi 3B+ computer, three cameras, and two DC motors. (a) Front view of camera system with inset of angle marker used in the platform angle detection mechanism. The software adds a red rectangle to show detection of the marker. (b) Side view of camera system. (c) Overall view of cameras, tripod, and subject. Cameras are at head level.

processor, cameras, and angle rotation apparatus, to record needed videos for subsequent image processing used in gait speed, standing balance, and 5TSS calculations.

Figure 4 displays images of the constructed CBMT device, complete with Raspberry Pi, cameras, and motors. The platform's physical dimensions are 13 cm × 19 cm × 36 cm (excluding the tripod). The gap between the left and right cameras is set at 273.5 mm to enhance distance accuracy. A marker symbol, located in the view area of the cameras, helps calculate the subject's overall angle, regardless of the platform rotation angle (see **Figure 4** inset). The CBMT system employs template matching to detect the marker and adjust the platform angle left/right or up/down [21]. For experiments requiring a 180° rotation of the platform, multiple markers can be used.

3. Software design and implementation

3.1 Software in RPi

The block diagram in **Figure 3** presents the internal functions of the hardware and software of the CBMT system. The Raspberry Pi block incorporates Thonny, an integrated development environment, facilitating Python script editing and execution. The graphical user interfaces (GUIs) are also developed as Python scripts, utilizing the tkinter (Tk interface) module. To enable cloud storage and easy accessibility of data online, Rclone, a command line program, is utilized [22].

In our design, we have installed Virtual Network Computing (VNC) Server and VNC Viewer in the Raspberry Pi, setting up a graphical desktop-sharing system. This setup allows the RPi Desktop to be viewed remotely on a cell phone screen through

VNC Viewer [23]. The cell phone serves as the interface, remotely controlling the CBMT system and collecting data, which is also saved in Dropbox and Google Drive. The image and data processing take place within the Raspberry Pi, with the cell phone only showing the results. The cell phone has access to all data, images, and video files on the Raspberry Pi.

For the multifunctional software of the proposed CBMT system, we utilized various tools, including Python3 programming language (version 3.7.3), OpenCV (cv2) library for computer vision (version 3.4.6), Imutils image processing functions (version 0.5.3), Channel and Spatial Reliability Tracking (CSRT) function from within OpenCV, tkinter module for GUI creation, Thonny development environment on the RPi, and VNC Server and Viewer on the RPi and cell phone respectively, Rclone command line program, and RPi.GPIO library for General Purpose Input–Output through the pins on the Raspberry Pi. The software overview is described in **Figure 5**, illustrating the steps and main equations involved in conducting gait speed, standing balance, and 5TSS tests. All the Python scripts written for this device are publicly available on Zenodo [24].

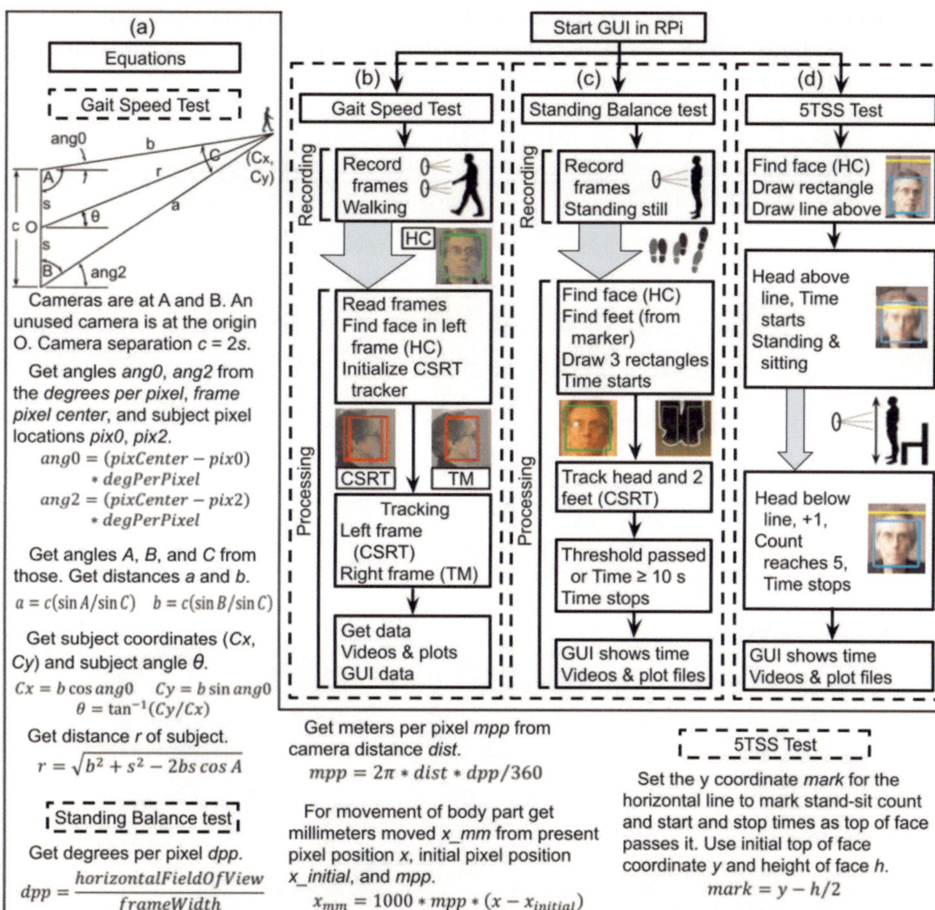

Figure 5.
(a) Equations for each of the three SPPB tests, (b) gait speed test flowchart, (c) standing balance test flowchart, and (d) 5TSS test flowchart. In the camera-based tracking system three types of object recognition are utilized: template matching (TM), Haar cascades (HC), and Channel and Spatial Reliability Tracking (CSRT).

3.2 Cell phone and graphical user interfaces

To establish a connection between the RPi and the cell phone for remote control of the CBMT system, Virtual Network Computing is utilized through the cell phone's Wi-Fi hotspot. The RPi's GUIs, including the control GUI and camera GUI, are displayed on the cell phone screen, and both GUIs are accessed with touch commands. Essentially, the cell phone acts as the remote monitor, keyboard, mouse, and router for the RPi.

In **Figure 6(a)**, the designed control GUI is presented, featuring a menu with options for gait speed, standing balance, and 5TSS tests. The control GUI includes buttons for running tests, displaying results on the phone, and saving data in comma-separated values (CSV) files. For each test, essential data such as the doctor's name, patient's ID, time stamp, and elapsed time are recorded. Additionally, for the gait speed test, supplementary data is kept for walking distance, walking distance error, average speed, and average of errors in a series of runs. The GUI also includes a button for syncing data to Google Drive and Dropbox, facilitating easy access to result files on those cloud platforms. Help buttons are available for user assistance, and the results are displayed in the control GUI while also being saved in CSV files and sent to the cloud with a single button press. Moreover, buttons are provided to delete video and plot files, and data for runs.

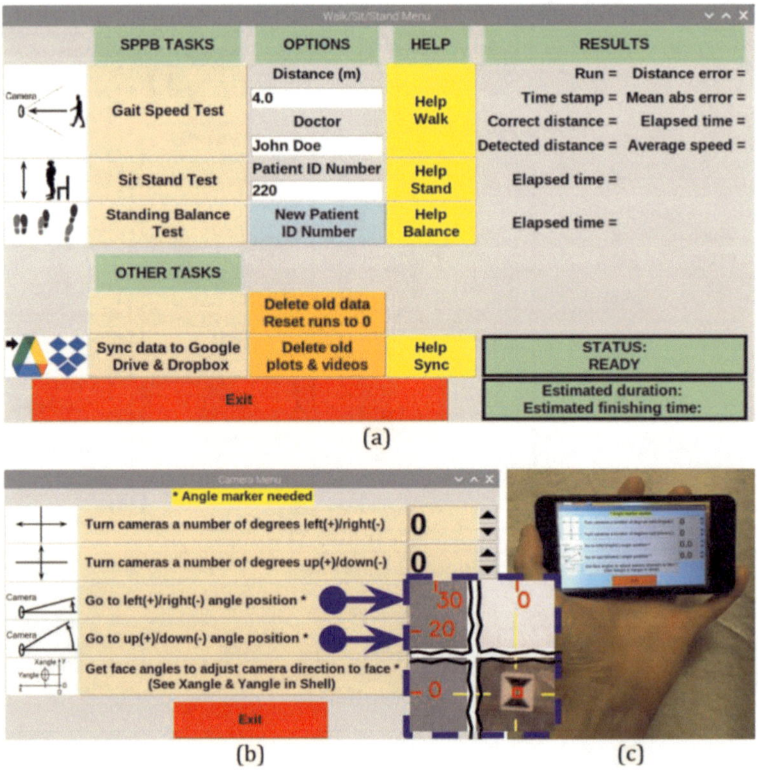

Figure 6.
(a) The control GUI displaying menu options for the three SPPB tests. (b) The camera GUI, with an inset showing degree axes on the left and top, showcasing the cameras centered horizontally and vertically on the angle marker. (c) The case of a cell phone being used for control.

In **Figure 6(b)**, the camera GUI has commands to operate the motors to adjust the angles of the camera platform. The first and second buttons allow the platform to turn a certain amount according to the number entered. The third and fourth buttons turn the platform's direction to a specified angle relative to the marker. Horizontal and vertical angle axes with the marker are displayed (see **Figure 6** inset). This setup process is crucial for preparing the system to commence a new experiment. **Figure 6(c)** illustrates the camera GUI when utilizing a cell phone for system control.

3.3 Object recognition techniques

This work utilizes three distinct object recognition (object detection) techniques: template matching (TM), Haar cascades (HC), and Channel and Spatial Reliability Tracking.

The template matching algorithm needs only one image as a template to search for a similar pattern of the same size and orientation within a larger image. A squared difference correlation algorithm is used to locate the best match. The grayscale template (T) is compared with all possible positions in a grayscale image (I). For a possible position in (I) with upper left corner pixel at (x, y), in Eq. (1) it obtains a sum $R(x, y)$ of squared differences of pixel values using the TM_SQDIFF method in the OpenCV function matchTemplate() [25, 26].

$$R(x,y) = \sum_{x',y'} (T(x',y') - I(x+x',y+y'))^2 \tag{1}$$

The pixel location (x, y) where the $R(x, y)$ sum is the minimum represents the best match for the template location in the image [25, 26]. This technique is applied to locate the angle marker in **Figure 4(a)**, aiding in determining the camera platform's angle in relation to the marker. Additionally, template matching is used to locate the subject's head in the right frame during detected walking path calculation in a gait speed test.

The second technique, Haar cascades, enables searches for more general patterns, such as human faces. It employs machine learning to create Haar cascade XML classifier files, which contain the necessary training for classification. The cv2 module's methods allow the identification of desired objects, such as faces, at different scales with the same orientation [27, 28]. Training involves using numerous known positive images with faces and known negative images without faces [28]. Haar features, simple classifiers, are applied to the images, which involve summing the intensities of adjacent rectangular areas and taking their differences. Integral images simplify these calculations by enabling fast summations of pixel values in rectangles of any size [29]. To create more complex classifiers, a machine learning process called AdaBoost is employed, which applies weights and adjusts them until they accurately classify each pixel according to its possibility of being a face location, considering different face sizes [27, 30]. These weights are used in the XML file model to look at new images and locate face positions. The term "cascades" refers to the multiple filtering passes that progressively eliminate large areas of the image and proceed with increased features and accuracy until the most probable face location is found [27].

The third technique, Channel and Spatial Reliability Tracking is a C++ implementation of the CSR-DCF (Channel and Spatial Reliability of Discriminative Correlation Filter) tracking algorithm within the OpenCV library [31]. The CSRT tracker trains the compressed features of Histogram of Oriented Gradients (HOG) and color names (colornames) using a correlation filter [32]. Compressed features in neural networks refer to the resulting dimensions after removing unnecessary features [33]. HOG represents the gradients of light intensity and their orientations, being higher at edges [34]. Colornames is a mapping from pixel values to color names, utilized to obtain features for color matching [35]. The CSRT tracking is initiated by selecting a region containing the subject's face in a video frame. In subsequent frames, a search is conducted near the same region by identifying features of color patches with their edges from the object and its surrounding region. This targeted approach accelerates the process by avoiding a search of the entire frame [32].

3.4 SPPB tests

In our implementation, we have employed three algorithms, each utilizing different object recognition techniques, to measure all SPPB tests. The flowcharts corresponding to these algorithms are presented in **Figure 5**. With these three tests, three types of motion are being tracked: 1) walking motion (gait speed), 2) small motions in two directions (standing balance), and 3) up and down motion (5 times sit-stand).

3.4.1 Gait speed test

First stage: recording videos: the gait speed test is recorded using both left and right cameras. At the beginning of the recording stage, the subject faces the cameras and starts walking along a planned path. This movement is simultaneously recorded by the two cameras, generating two video files.

Second stage: processing video frames for gait speed and walking path calculation: frames from the two video files are processed. Initially in the left frame, the number of pixels both horizontally and vertically, is increased by three times. This is to enable accurate face detection at distances of 6 or 7.2 m by Haar cascades. When the face is found, the CSRT tracker is initialized with the face location and Haar cascades is no longer needed. The CSRT tracker maintains the subject's pixel location in subsequent left video frames, and no increased resolution is required for CSRT.

For each frame pair from initialization to the end, the subject's pixel location in the right frame is determined by using a template from the left frame and finding a match in the right frame through template matching. The left and right cameras are angled in a nonparallel direction toward a point 4 m away, ensuring that the subject remains relatively more centered in each frame. Pixel adjustments are made for the subject's pixel position in the left and right frames to correct for the nonparallel camera setup.

The subject's pixel locations from both cameras undergo filtering using customized median and moving average filters. These filters are applied minimally to remove noise and straighten the trajectory from any side-to-side swaying, resulting in a more accurate calculation of the distance moved.

In previous work [21], the triangulation technique was employed to obtain polar coordinate positions from subject pixel position pairs (see **Figure 5(a)**). To reduce depth inaccuracies in the gait speed calculation, the system was calibrated using Eqs. (2) and (3), where cubic equations were fitted to data obtained from 64 locations on an 8-by-8 rectangular grid covering the experiment area.

Eq. (2) involves creating a function, *depth_f*, using Python's *interp2d* function for 2D interpolation, with lists X and Y representing rectangular coordinates for detected depths and angles, and R representing correct depths corresponding to the other two lists.

$$depth\_f = interp2d(X, Y, R, 'cubic'), \tag{2}$$

Eq. (3) calculates the correct depth, *dep*, for a single case using rectangular coordinates x and y for a detected depth and angle.

$$dep = depth\_f(x, y), \tag{3}$$

3.4.2 Standing balance test

To assess standing balance, the center camera records the subject while they remain stationary. In the standing balance test the subject must stand with feet together in three positions: side-by-side, semi-tandem, and tandem, each for a

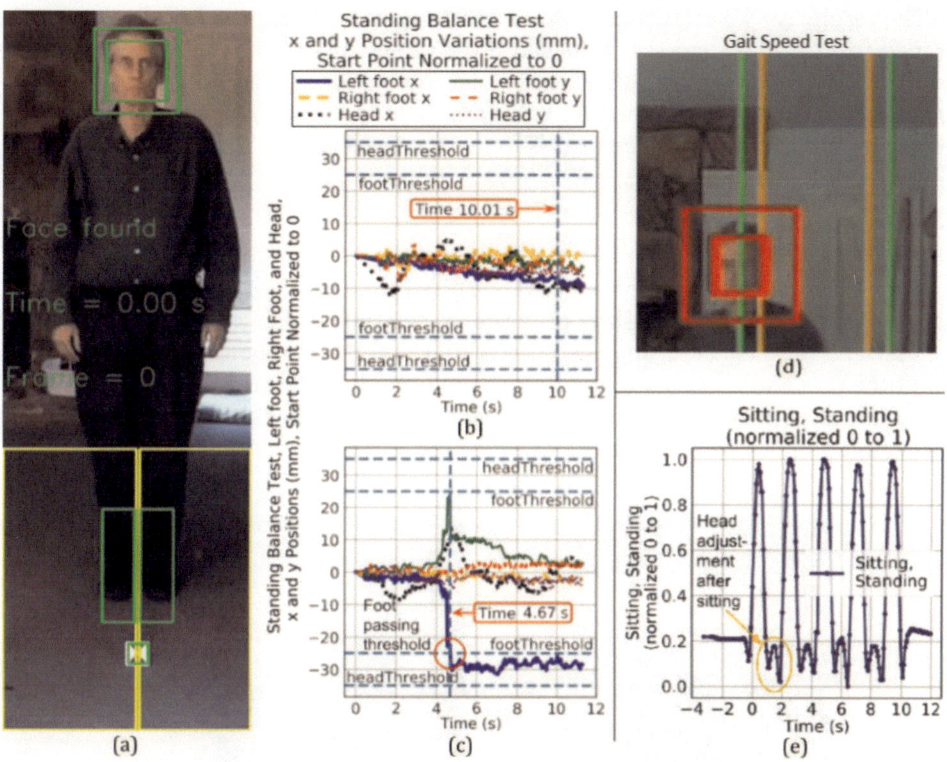

Figure 7.
Measurement from various tests. (a) Video footage displays three green tracking boxes: the head (outer green box) and feet (lower green boxes). Additionally, two yellow independent frames at the bottom are for the feet trackers, and a small green box surrounds a marker at the bottom helping determine the initial location of the feet. (b) A plot representing a successful stand lasting over 10 seconds during the standing balance test. (c) A plot showing a failing stand of 4.67 seconds because the left foot moved about 30 mm to the left. This is evident from the thick blue L foot x *curve, which crosses the foot threshold at −25 mm (marked with a red circle). (d) During the gait speed test, the subject is walking toward the camera. Vertical green lines mark the maximum and minimum horizontal pixel values of the tracked head. Vertical orange lines indicate the start and finish pixel positions when the subject is in motion, making the start and finish times more distinct. (e) The 5 times sit-stand test is displayed with the sitting and standing data normalized from 0 to 1.*

duration of 10 seconds without moving their feet or head [8]. There is an option to ignore head movements during the tests.

To prepare for the test, the floor is measured from the camera. There is masking tape at 3.5 m for the feet's location marker, and at 4 m for the standing position. The cameras are directed toward the person's crotch. To center the subject in the view and increase the frame rate, pixels are automatically trimmed from the left and right sides of the frame. From the video recording, the developed code utilizes Haar cascades to detect the subject's face. By using template matching, an angle marker is located to determine the relative positions of the left and right feet (see **Figure 7(a)**).

For the tandem position, the subject stands with feet slightly angled toward the side of the forward foot, ensuring both feet are visible to the camera. The analysis involves three CSRT tracking boxes for the head [36], left foot, and right foot, as shown in **Figure 5(c)**. Each foot tracker operates independently within its designated yellow frame (**Figure 7(a)**) to avoid interference from the other foot. The software monitors movements of these body parts and triggers alerts if any of them exceed the distance thresholds set (feet: >25 mm, head: >35 mm, see **Figure 7**). These movements are represented by horizontal x-type and vertical y-type curve labels.

The results can be classified as Pass or Fail scores or given numerical scores within a range (e.g., 1–10). Success is achieved if the subject maintains the required posture for 10 or more seconds, and failure or partial success if the duration is less than 10 seconds. The average frame rate is 22 frames per second (fps), and the time resolution for standing balance tests is under 0.055 seconds.

3.4.3 5 times sit-stand test

The 5TSS test algorithm measures the elapsed time for the subject to perform 5 stand-sit cycles. **Figure 8** displays prompts for the test, which can be viewed by both the user and the subject. Using Haar cascades, the software identifies the level of the top of the face and increments the count each time the subject completes a stand-sit cycle. The test operates at an average frame rate of 25 frames per second (fps), with a time resolution of under 0.050 seconds for 5TSS tests.

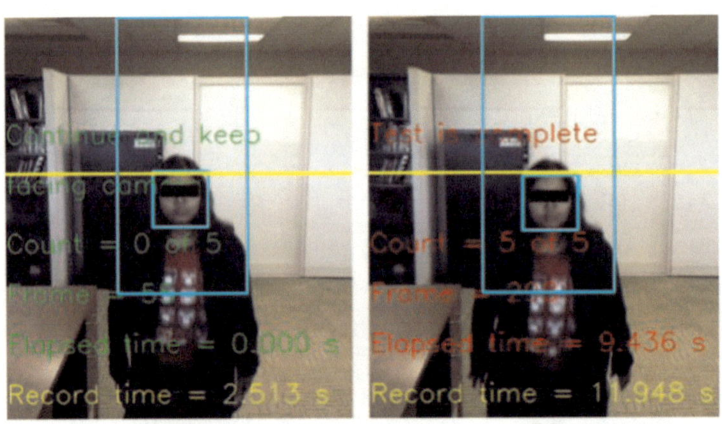

(a) (b)

Figure 8.
The 5 times sit-stand test. (a) The top of subject's head box rises above the yellow line, initiating the timer. (b) The top of subject's head box falls below the yellow line for the fifth time, leading to the timer's stop at 9.436 s.

4. Experimental results, SPPB

4.1 Human subject population and testing environments

The experiments were conducted on two groups of healthy subjects without cancer: young and old adults. The CBMT platform (shown in **Figure 4**) was utilized to perform in vivo experiments on eight human volunteers. The first group comprised four individuals aged 60–95, consisting of three males and one female, all with light complexions. The second group included four individuals aged 22–42, comprising three males and one female, with three having light complexions and one with a dark complexion. All SPPB tests were conducted using the same CBMT system.

The CBMT system is powered by a battery power bank, ensuring continuous operation for up to 8 hours. It can be initialized and ready to execute tests in less than 45 seconds after being turned on. The simplicity of using the CBMT system is highlighted by the fact that the operator also served as the subject in some of the experiments.

4.2 Results, gait speed test

Figure 9 presents the distances walked and walking speeds observed during the gait speed test using the CBMT prototype. In **Figure 9(a)**, the two vertical light blue lines indicate the start and end of timing for the walk, automatically generated by the software through angle analysis over time. Start and finish points are defined more distinctly by recording them with the subject in motion. Also see **Figure 7(d)** and its explanation in the caption. The smooth and steady curve in **Figure 9(a)** suggests that the subject had good balance and maintained a consistent pace without faltering or pausing. Minimal noise filtering was applied to the data.

During the test, the subject walked at a normal pace, taking approximately 4 seconds to cover about 4 meters. The corresponding instantaneous walking speed is depicted in **Figure 9(b)**, showing a gradual increase in speed until the middle of the plot, followed by a decrease toward the end. The red horizontal line represents the average speed (Ave Speed 1), calculated by dividing the total distance walked by the total time.

Additionally, the orange horizontal dotted line corresponds to the average speed (Ave Speed 2), calculated by determining the total distance covered based on the green area under the curve. This distance is then divided by the total time, as shown in Eq. (4) where t denotes time, s represents speed, and the subscripts i and $i + 1$ denote indices of time and speed values.

$$AveSpeed2 = \frac{\sum_{i=1}^{n-1}[(t_{i+1} - t_i)(s_i + s_{i+1})/2]}{totaltime}, \tag{4}$$

Furthermore, **Figure 9(c)** displays a polar plot of a walking crosswise trajectory before filtering out the left-and-right jitters and before applying 2D interpolation for depth correction. **Figure 9(d)** illustrates the trajectory after these adjustments.

4.3 Results, standing balance test

The results for the standing balance tests are presented in the graphs of **Figures 7 (b)** and **(c)**. These plots show the horizontal (x-type label) and vertical (y-type label)

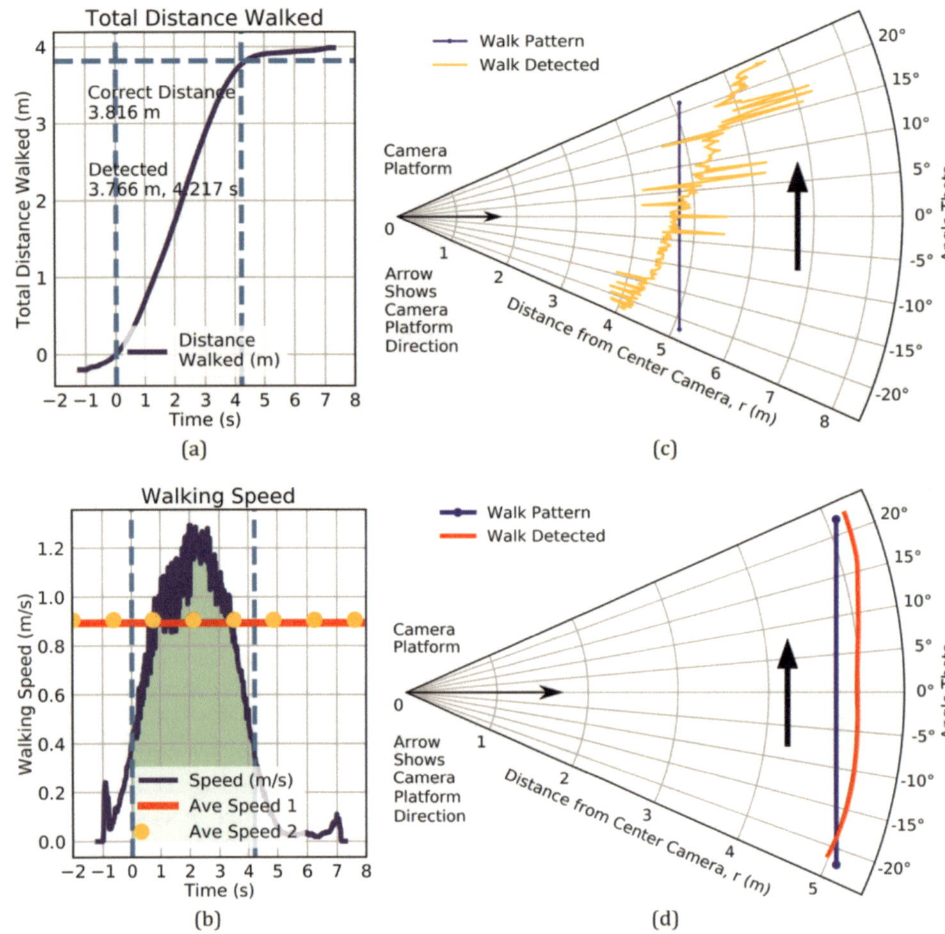

Figure 9.
(a) Total distance walked with blue vertical lines marking the start and finish points; (b) walking speed data, where Ave Speed 1 is indicated by the red line and Ave Speed 2 is shown by orange dots; (c) unfiltered and uncorrected orange walking trajectory; and (d) final adjusted red walking trajectory.

movements in millimeters for the head, left foot, and right foot of each subject. In **Figure 7(b)**, the subject successfully maintains a steady stance for over 10 seconds. However, in **Figure 7(c)**, the subject exhibits slight movement to the left, and thus, is credited with standing still for 4.67 seconds. To account for the small jitter detected even when standing still, a distance threshold is applied to determine whether the subject has moved or remained steady.

4.4 Results, 5 times sit-stand test

Figure 7(e) displays a normalized plot showing the height of the top of the subject's face as a function of time during a 5TSS test. The subject did the test smoothly, evident from the five smooth rises and falls of the evenly spaced curve. After each time sitting, there is a natural head position adjustment, due to leaning back in the chair, which is indicated by a brown oval marking one of these adjustments in the lower part of the plot.

4.5 Human subject population results

The CBMT platform was utilized to gather data for the gait speed, standing balance, and 5TSS tests, performed in multiple runs to assess the system's precision and consistency (**Table 1** and **Figure 10**). **Table 1** presents the measured results for all the tests. A total of 25 runs were conducted for gait speed walking, with 17 runs performed crosswise to the cameras and eight runs toward the cameras. The first column provides details on frame rates and pixel resolutions for each test. The second column displays the average gait speed ranges for all 25 runs, while the third column shows the gait speed resolutions, which were obtained by multiplying the absolute values of the decimal errors by the average speeds. The distances walked and time ranges for the gait speed tests are shown in the second-to-last column. However, there is no walking involved in the standing balance or 5TSS tests. The last data column

Type of test	Frame rate (frames/s), resolution (pixels)	Average speed range (m/s)	Gait speed resolution (m/s)	Distance (m)/time range (s)	Average accuracy
Gait speed, Crosswise, 8 subjects, 17 runs, **Figure 9**.	53, 640 × 96 × 2 frames = 122,880	0.191–1.259	<0.127	2.832–3.836/2.75–20.06	Distance accuracy: 95.28%
Gait speed, Radial, 1 subject, 8 runs, **Figure 7 (d)**.	39, 640 × 144 × 2 frames = 184,320	0.136–1.371	<0.067	3.866/2.82–28.35	Distance accuracy: 97.47%
Standing balance (3 types), 8 subjects, 18 runs, **Figures 7(a)–(c)**	22, 304 × 848 = 257,792	N/A	N/A	0.0/0.75–10.09	Time accuracy: 97.25%
5 Times sit-stand, 7 subjects, 11 runs, **Figure 7 (e)**	25, 400 × 480 = 192,000	N/A	N/A	0.0/6.93–16.94	Time accuracy: 99.10%

Table 1.
Results of 54 test runs with eight volunteers performed by CBMT system.

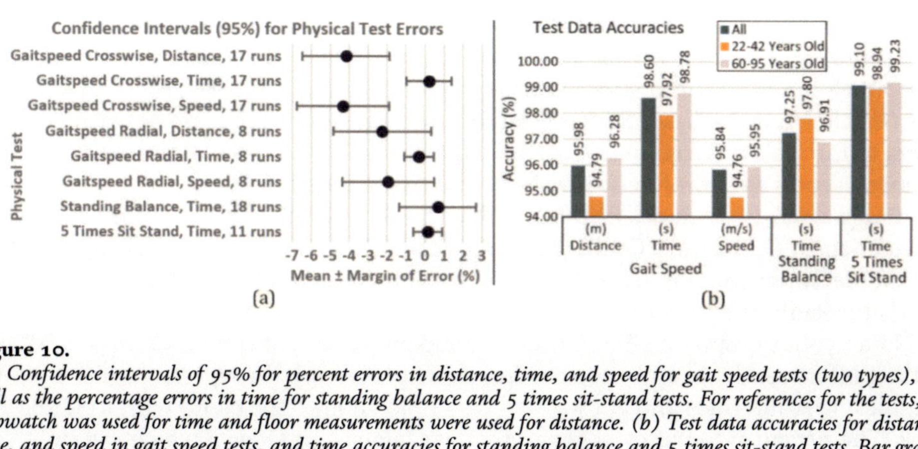

Figure 10.
(a) Confidence intervals of 95% for percent errors in distance, time, and speed for gait speed tests (two types), as well as the percentage errors in time for standing balance and 5 times sit-stand tests. For references for the tests, a stopwatch was used for time and floor measurements were used for distance. (b) Test data accuracies for distance, time, and speed in gait speed tests, and time accuracies for standing balance and 5 times sit-stand tests. Bar graph categories are age groups including younger, older, and all.

gives the measured average distance accuracies (>95%) for the gait speed tests and time accuracies (>97%) for the other tests.

In **Figure 10(a)**, the 95% confidence intervals for errors in distance, time, and average speed for gait speed tests, as well as time errors for standing balance and 5 times sit-stand tests, are presented. In **Figure 10(b)**, the accuracies of test data for both younger and older age groups, as well as overall, are depicted. The differences between age groups are relatively small.

5. Discussion

In this section, we compare the performance of the proposed CBMT technology, which enables gait speed, standing balance, and 5TSS tests, with various commercialized technologies and state-of-the-art published works (comparison, **Table 2**). The results are summarized in **Table 2**, focusing on the accuracies or errors achieved by different technologies for the three tests.

Several technologies, such as MetaWear CPro sensor, Unsupervised Screening System, and APDM ML system/Opal sensors, employ Inertial Measurement Units (IMUs) or accelerometers. While some wearable accelerometer-based devices may provide relatively accurate displacement after calibration, they suffer from displacement error accumulation, which can impact overall accuracy. The accuracy of displacement measurement from IMU devices, based on acceleration and its conversion to gait speed, is complex and varies based on sensor placement, subjects' ages, and IMU performance.

Other technologies like Gaitspeedometers utilize ultrasound. Time-of-flight sensors may use infrared, lasers, or ultrasound to measure subject depth. The Kinect device can detect body postures and utilizes both infrared and visible light. However, devices with infrared cameras can be expensive, and setups involving lasers or ultrasound for light barriers are less mobile.

In **Table 2** Accuracy is calculated using the formula in Eq. (5) [38]:

$$Accuracy = \left(100 - \left|\frac{TrueValue - MeasuredValue}{TrueValue}\right| \times 100\right)\%. \qquad (5)$$

Tech refers to Technology. **IMU** stands for Inertial Measurement Unit. **APDM ML** refers to Ambulatory Parkinson's Disease Monitoring MobilityLab/Opal sensors. **LB** represents Light Barrier Sensor. **aTUG** stands for Ambient Timed "Up & Go" chair, which measures TUG automatically using an infrared Light Barrier, four Force Sensors (on chair), and a Laser Range-Scanner. **Acc/Err** denotes Accuracy/Error. **ICC** refers to Intra-class Correlation Coefficients, where the GAITRite pressure sensor walkway serves as the reference system. **Ave Acc** stands for Average Accuracy. **Coverage area** also includes distance walked on a treadmill. **LB CC** refers to Light Barrier Sensor Correlation Coefficient (between it and stopwatch). **IMU CC** denotes Inertia Measurement Unit Correlation Coefficient (between it and stopwatch). **RFID** stands for Radio-frequency Identification.

Notably, our proposed CBMT system is the only one that covers all three SPPB tests. It offers results in real-time or after short processing through an intuitive graphical user interface and allows easy data uploading to the cloud with a simple click. The system automatically generates data, plots, and videos, which can be effortlessly removed when no longer needed.

Parameters	**Name/Tech/Company/Reference**					
	Gaitspeedometers 4 Types [37]	**MetaWear CPro sensor [38]**	**Microsoft Kinect V2 [39]**	**Unsupervised Screening System [40]**	**APDM ML system (version 2) [41, 42]**	**This Work (54 test runs)**
Sensor type	Laser, infrared, ultrasonic	IMU	Camera, infrared	LB, aTUG, IMU, camera	IMU	Visible light cameras
Gait speed test	YES	YES	YES	NO	YES	YES
Distance Acc/Err (%)	–	97.57% & 89.3% Ave Acc	–	N/A	–	95.98% Ave Acc
Gait speed Acc/Err (%)	98.3%, 86.1%, 98.6%, 95.9% Ave Acc	97.57% & 89.3% Ave Acc	–	N/A	ICC = 0.928	95.84% Ave Acc
Time Acc/Err (%)	98.3%, 86.1%, 98.6%, 95.9% Ave Acc	99.01% Ave Acc	–	N/A	–	98.59% Ave Acc
Coverage area (m, m^2)	4 m	~38 m	3 m	~3.5 m	6 m × 0.6 m	2 to 7.2 m, 60° view
Standing balance test	NO	NO	NO	NO	YES	YES
Time (s) Acc/Err (%)	N/A	N/A	N/A	N/A	–	97.25% Ave Acc
5 Times sit-stand test	NO	NO	NO	YES	NO	YES
Time (s) Acc/Err (%)	N/A	N/A	N/A	LB CC 0.73 IMU CC 0.87	N/A	99.10% Ave Acc
Data handling	Phone displays, microcontrollers	Bluetooth, phone	USB	RFID, cables	wireless	Wi-Fi, Cloud storage
Processor	Microcontrollers	Android phone, app	Computer, Kinect V2	Computer	Laptop	Raspberry Pi

Table 2.
Technologies used for gait speed, standing balance, and 5 times sit-stand tests.

6. Conclusion

We have successfully designed and implemented a camera-based system capable of evaluating all three SPPB tests for the physical assessment of cancer patients, including gait speed, standing balance, and 5TSS tests. Our CBMT software comprises codes and algorithms with GUIs that process images captured by the three cameras to assess SPPB. Additionally, we have incorporated remote control functionality via smartphones, enabling easy access to GUIs and automatic data collection for evaluating a patient's physical status. The system efficiently stores data on the Raspberry Pi and allows for cloud uploading.

Through in vivo experiments involving a human subject population of eight volunteers, including four older individuals aged 60–95 (both male and female), we have obtained promising results. For gait speed measurement tests, we achieved an average accuracy of over 95% for distance traveled and average gait speed. Furthermore, the average time accuracies for standing balance and 5TSS tests exceeded 97%. These outcomes demonstrate the system's effectiveness in providing reliable and precise evaluations of physical performance. Such data can play a vital role in personalizing cancer treatment strategies, particularly for older adults.

Notes

This current chapter is based on and is a continuation of the previously published work, "Camera-Based Short Physical Performance Battery and Timed Up and Go Assessment for Older Adults with Cancer," and one other paper, "Camera-Based Human Gait Speed Monitoring and Tracking for Performance Assessment of Elderly Patients with Cancer [20, 21]."

Author details

Larry Duncan[1], Shaotong Zhu[2], Mackenzi Pergolotti[3,4], Smith Giri[1], Hoda Salsabili[5], Miad Faezipour[5], Sarah Ostadabbas[2] and S. Abdollah Mirbozorgi[1]*

1 The University of Alabama at Birmingham, Birmingham, AL, USA

2 Northeastern University, Boston, MA, USA

3 University of North Carolina, Chapel Hill, NC, USA

4 Select Medical, Mechanicsburg, PA, USA

5 Purdue University, West Lafayette, IN, USA

*Address all correspondence to: samir@uab.edu

References

[1] Smith BD, Smith GL, Hurria A, Hortobagyi GN, Buchholz TA. Future of cancer incidence in the United States: Burdens upon an aging, changing nation. Journal of Clinical Oncology. 2009; **27**(17):2758-2765

[2] Miller KD, Siegel RL, Lin CC, Mariotto AB, Kramer JL, Rowland JH, et al. Cancer treatment and survivorship statistics, 2016. CA: A Cancer Journal for Clinicians. 2016;**66**(4):271-289

[3] Wu HS, Harden JK. Symptom burden and quality of life in survivorship: A review of the literature. Cancer Nursing. 2015;**38**(1):E29-E54

[4] Neo J, Fettes L, Gao W, Higginson IJ, Maddocks M. Disability in activities of daily living among adults with cancer: A systematic review and meta-analysis. Cancer Treatment Reviews. 2017;**61**: 94-106

[5] Pergolotti M, Deal AM, Williams GR, Bryant AL, McCarthy L, Nyrop KA, et al. Older adults with cancer: A randomized controlled trial of occupational and physical therapy. Journal of the American Geriatrics Society. 2019;**67**(5): 953-960

[6] Mohile SG, Dale W, Somerfield MR, Schonberg MA, Boyd CM, Burhenn PS, et al. Practical assessment and management of vulnerabilities in older patients receiving chemotherapy: ASCO guideline for geriatric oncology. Journal of Clinical Oncology. 2018;**36**(22): 2326-2347

[7] Giri S, Al-Obaidi M, Harmon C, Dai C, Smith CY, Gbolahan OB, et al. Impact of geriatric assessment (GA) based frailty index (CARE-Frailty Index) on mortality and treatment-related toxicity among older adults with gastrointestinal (GI) malignancies. Journal of Clinical Oncology. 2021;**39** (15_suppl):12046-12046

[8] Marsh AP, Wrights AP, Haakonssen EH, Dobrosielski MA, Chmelo EA, Barnard RT, et al. The virtual short physical performance battery. The Journals of Gerontology. Series A, Biological Sciences and Medical Sciences. 2015;**70**(10):1233-1241

[9] Dale W, Williams GR, MacKenzie AR, Soto-Perez-de-Celis E, Maggiore RJ, Merrill JK, et al. How is geriatric assessment used in clinical practice for older adults with cancer? A survey of cancer providers by the American Society of Clinical Oncology. JCO Oncology Practice. 2021;**17**(6): 336-344

[10] Ezzatvar Y, Ramírez-Vélez R, Sáez de Asteasu ML, Martínez-Velilla N, Zambom-Ferraresi F, Izquierdo M, et al. Physical function and all-cause mortality in older adults diagnosed with cancer: A systematic review and meta-analysis. The Journals of Gerontology. Series A, Biological Sciences and Medical Sciences. 2021;**76**(8):1447-1453

[11] Seeed Technology Co.,Ltd. Raspberry Pi Camera Module V2 [Internet]. Seeed Studio; 2022. Available from: https://www.seeedstudio.com/Raspberry-Pi-Camera-Module-V2.html

[12] Wikipedia. Laser [Internet]. Wikipedia, the free encyclopedia; 2021. Available from: https://en.wikipedia.org/wiki/Laser

[13] Wikipedia. Kinect [Internet]. Wikipedia, the free encyclopedia; 2021. Available from: https://en.wikipedia.org/wiki/Kinect

[14] Kong K, Tomizuka M. A gait monitoring system based on air pressure sensors embedded in a shoe. IEEE/ASME Transactions on Mechatronics. 2009; **14**(3):358-370

[15] SainSmart. Ultrasonic Ranging Detector Mod HC-SR04 Distance Sensor [Internet]. SainSmart.com; 2023. Available from: https://www.sainsmart.com/products/ultrasonic-ranging-detector-mod-hc-sr04-distance-sensor

[16] SparkFun Electronics. SparkFun 9DoF IMU Breakout-LSM9DS1-SEN-13284-SparkFun Electronics [Internet]. SparkFun Electronics. Available from: https://www.sparkfun.com/products/retired/13284 [Accessed: June 14, 2023]

[17] Ostadabbas S, Sebkhi N, Zhang M, Rahim S, Anderson LJ, Lee FEH, et al. A vision-based respiration monitoring system for passive airway resistance estimation. IEEE Transactions on Biomedical Engineering. 2016;**63**(9): 1904-1913

[18] Wang Z, Mirbozorgi SA, Ghovanloo M. An automated behavior analysis system for freely moving rodents using depth image. Medical & Biological Engineering & Computing. 2018;**56**(10):1807-1821

[19] Aloudat M, Faezipour M, El-Sayed A. Automated vision-based high intraocular pressure detection using frontal eye images. IEEE Journal of Translational Engineering in Health and Medicine. 2019;**7**:3800113

[20] Duncan L, Zhu S, Pergolotti M, Giri S, Salsabili H, Faezipour M, et al. Camera-based short physical performance battery and timed up and go assessment for older adults with cancer. IEEE Transactions on Biomedical Engineering. Sep 2023;**70**(9):2529-2539

[21] Duncan L, Gulati P, Giri S, Ostadabbas S, Abdollah Mirbozorgi S. Camera-based human gait speed monitoring and tracking for performance assessment of elderly patients with cancer. In: Annual International Conference of the IEEE Engineering in Medicine and Biology Society 2021. 2021. pp. 3522-3525

[22] Jarrod. Project: RaspberryPi Google Drive Sync [Internet]. Jarrod's Tech; 2018. Available from: https://jarrodstech.net/project-raspberrypi-google-drive-sync/ [Accessed: June 14, 2023]

[23] Srinivasulu B, Kuamari DS, Srinivas G, Rao GJ. VNC viewer authentication and security using remote frame buffer protocol. International Journal of Engineering Research and Applications. 2013;**3**(6):1007-1011

[24] Duncan LD. Raspberry Pi Camera Tracking [Internet]. Zenodo; 2023. Available from: https://zenodo.org/record/7644881 [Accessed: June 12, 2023]

[25] OpenCV. Template Matching [Internet]. OpenCV, Open Source Computer Vision; 2019. Available from: https://docs.opencv.org/3.4.6/de/da9/tutorial_template_matching.html

[26] Basulto-Lantsova A, Padilla-Medina JA, Perez-Pinal FJ, Barranco-Gutierrez AI. Performance comparative of OpenCV template matching method on Jetson TX2 and Jetson Nano developer kits. In: 2020 10th Annual Computing and Communication Workshop and Conference (CCWC) [Internet]. Las Vegas, NV, USA: IEEE; 2020. pp. 0812-0816. Available from: https://ieeexplore.ieee.org/document/9031166/ [Accessed: June 12, 2023]

[27] Viola P, Jones M. Rapid object detection using a boosted cascade of simple features. In: Proceedings of the

2001 IEEE Computer Society Conference on Computer Vision and Pattern Recognition CVPR 2001 [Internet]. Kauai, HI, USA: IEEE Computer Society; 2001. pp. I-511-I–518. Available from: http://ieeexplore.ieee.org/document/990517/ [Accessed: June 12, 2023]

[28] OpenCV. Face Detection Using Haar Cascades [Internet]. OpenCV; 2019. Available from: https://docs.opencv.org/3.4.6/d7/d8b/tutorial_py_face_detection.html#gsc.tab=0

[29] Rahmad C, Asmara RA, Putra DRH, Dharma I, Darmono H, Muhiqqin I. Comparison of Viola-Jones Haar cascade classifier and histogram of oriented gradients (HOG) for face detection. IOP Conference Series: Materials Science and Engineering. 2020;**732**(1):012038

[30] Wang R. AdaBoost for feature selection, classification and its relation with SVM, a review. Physics Procedia. 2012;**25**:800-807

[31] Farkhodov K, Lee SH, Kwon KR. Object tracking using CSRT tracker and RCNN. In: Proceedings of the 13th International Joint Conference on Biomedical Engineering Systems and Technologies [Internet]. Valletta, Malta: SCITEPRESS-Science and Technology Publications; 2020. pp. 209-212. Available from: http://www.scitepress.org/DigitalLibrary/Link.aspx?doi=10.5220/0009183802090212 [Accessed: June 12, 2023]

[32] Lukežič A, Vojíř T, Čehovin Zajc L, Matas J, Kristan M. Discriminative correlation filter tracker with channel and spatial reliability. International Journal of Computer Vision. 2018; **126**(7):671-688

[33] Taylor P, Griffiths N, Hall V, Xu Z, Mouzakitis A. Feature selection for supervised learning and compression. Applied Artificial Intelligence. 2022; **36**(1):2034293

[34] Dalal N, Triggs B. Histograms of oriented gradients for human detection. In: 2005 IEEE Computer Society Conference on Computer Vision and Pattern Recognition (CVPR'05) [Internet]. San Diego, CA, USA: IEEE; 2005. pp. 886-893. Available from: http://ieeexplore.ieee.org/document/1467360/ [Accessed: June 12, 2023]

[35] Van De Weijer J, Khan FS. An overview of color name applications in computer vision. In: Trémeau A, Schettini R, Tominaga S, editors. Computational Color Imaging [Internet]. Cham: Springer International Publishing; 2015. pp. 16-22. (Lecture Notes in Computer Science; vol. 9016). Available from: http://link.springer.com/10.1007/978-3-319-15979-9_2 [Accessed: June 12, 2023]

[36] Al-Rahayfeh A, Faezipour M. Eye tracking and head movement detection: A state-of-art survey. IEEE Journal of Translational Engineering in Health and Medicine. 2013;**1**:2100212

[37] Jung HW, Roh HC, Kim SW, Kim S, Kim M, Won CW. Cross-comparisons of gait speeds by automatic sensors and a stopwatch to provide converting formula between measuring modalities. Annals of Geriatric Medicine and Research. 2019;**23**(2):71-76

[38] Anwary AR, Yu H, Vassallo M. An automatic gait feature extraction method for identifying gait asymmetry using wearable sensors. Sensors (Basel). 2018; **18**(2):676

[39] Dubois A, Bihl T, Bresciani JP. Automating the timed up and go test using a depth camera. Sensors. 2017;**18**(2):14

[40] Fudickar S, Hellmers S, Lau S, Diekmann R, Bauer JM, Hein A.

Measurement system for unsupervised standardized assessment of timed "up & go" and five times sit to stand test in the community-A validity study. Sensors (Basel). 2020;**20**(10):2824

[41] Morris R, Stuart S, McBarron G, Fino PC, Mancini M, Curtze C. Validity of mobility lab (version 2) for gait assessment in young adults, older adults and Parkinson's disease. Physiological Measurement. 2019;**40**(9):095003

[42] Fino PC, Horak FB, El-Gohary M, Guidarelli C, Medysky ME, Nagle SJ, et al. Postural sway, falls, and self-reported neuropathy in aging female cancer survivors. Gait & Posture. 2019;**69**: 136-142

Chapter 4

Dynamic Balance in the Gait Cycle Prior to a 90° Turn in Individuals with Parkinson's Disease

Gordon Alderink, Cathy Harro, Lauren Hickox, David W. Zeitler, Dorothy Kilvington, Rebecca Prevost and Paige Pryson

Abstract

Parkinson's disease (PD), a prevalent neurodegenerative condition, is associated with fall-related injuries. Falls often occur during mobility tasks such as turning while walking. There is a paucity of research on the biomechanical etiology of falls, specifically, the control of dynamic balance during turns. The purpose of this study was to analyze dynamic stability, as measured by the margin of stability (MOS), during the gait cycle preceding a 90-degree turn during walking in persons with PD. Thirteen individuals with mild to moderate idiopathic PD and 10 healthy matched controls (CON) participated. Instrumented gait analysis was conducted during walking while performing 90-degree turns using the Plug-in Gait model and Vicon Nexus motion capture software. MOS variables at first double support, midstance, and second double support of the gait cycle preceding the turn were examined. The MOS variables and spatiotemporal gait parameters were compared between PD and CON using a multilevel mixed model ANOVA; post hoc analyses were conducted using two-sample t-tests. There were no differences in spatiotemporal gait parameters between groups. The PD group demonstrated significantly greater medio-lateral (M/L) MOS compared to CON for most variables. The changes seen in the M/L MOS in the PD group may reflect compensatory changes to increase dynamic stability during the gait cycle preceding a turn.

Keywords: gait, spatiotemporal parameters, center of mass, extrapolated center of mass, center of pressure, margin of stability

1. Introduction

Parkinson's disease (PD) is the second most common neurological condition in the world, affecting more than 10 million people [1]. This disorder results from the degeneration of dopaminergic neurons within the substantial nigra causing changes in nigrostriatal pathways in the basal ganglia [2]. The basal ganglia are involved with the control of voluntary movement, postural tone, the automaticity of postural control strategies, and efficient gait function. Parkinson's disease results in progressive gait and balance decline and difficulty adapting walking to varied task and environmental demands such as managing turns and obstacles [3–5]. Due to the effects of PD on the gait and postural control systems, all dynamic balance and walking activities require increased attention, which increases the cognitive load of locomotion [3]. Individuals with PD demonstrate increased gait variability and difficulty performing a second task while walking, which is also reflective of impaired balance and gait automaticity [3, 5]. Falls among the elderly population lead to the hospitalization of more than 800,00 Americans each year [6], and persons with PD are at even greater fall risk [7]. Falls during turns are very common in persons with PD and are eight times more likely to result in a hip fracture than other mechanisms of injury [8–10]. These injurious falls can force individuals with PD to become homebound, which reduces independence in their activities of daily living and participation in other meaningful activities.

Changing direction or turning while walking requires more proactive balance strategies and increased inter-limb coordination as compared to continuous forward walking [11]. Neural systems related to turning may be more vulnerable to functional impairments in individuals with PD [12]. The natural aging process results in gait changes including decreased gait velocity, stride length, single limb support time, and altered step width [13, 14]. In addition to these age-related changes, persons with PD also demonstrate PD-specific gait impairments including gait hypokinesia, shuffling gait with increased cadence and decreased stride length, reduced gait speed, and freezing of gait [5]. During walking and turning, persons with PD have impaired movement strategies [12], compromised head and neck stabilization, and abnormal muscle activation patterns [15]. Axial coordination is impaired during turns while walking as persons with PD display "en-bloc turning", observed as a simultaneous rotation of their head and upper trunk instead of a cranial to caudal sequential rotation that is observed in healthy individuals [12]. Previous research provides evidence that persons with PD approach turns more slowly, take longer to turn, take more steps, and perform turns that are wider and less accurate than healthy age-matched adults [8, 12, 16]. During a turn, a person's center of mass (COM) may be outside their base of support (BOS). Compared to healthy individuals during a walking turn, those with PD display a narrower BOS, which may lead to decreased dynamic stability and increased fall risk [8, 9]. Previous studies have examined the quality and sequencing of turning in persons with PD [8, 12, 17, 18], but research examining the biomechanical variables of dynamic balance during walking with turns is limited. Anticipatory postural strategies support dynamic stability during turns [12, 14]. At the time of this paper, no studies have analyzed the dynamic balance variables of the gait cycle just prior to a 90-degree turn. Given the motor planning impairments seen in persons with PD, these findings emphasize the necessity of continued research into the specific challenges related to the preparatory gait cycle directly before a turn.

The inverted pendulum model has been used to examine static human balance (**Figure 1**) [19]. It models the human body as a COM, hinged at the ankle joint, moving around a static BOS. In this model, the location of the center of pressure (COP) can be defined as the reaction to changes in the movements of the COM. The model has since been adapted to allow the representation of dynamic movement, like walking and turning, by utilizing additional variables associated with dynamic balance [19]. Among these variables is the COM-COP inclination angle, i.e. the spatial relationship between the COM and COP, which has been used to analyze gait stability during sharp turns [21]. The extrapolated center of mass (XCOM) is a measure that accounts for the horizontal velocity of the COM during walking, and the margin of stability (MOS) is the minimum distance from XCOM to the boundaries of the BOS

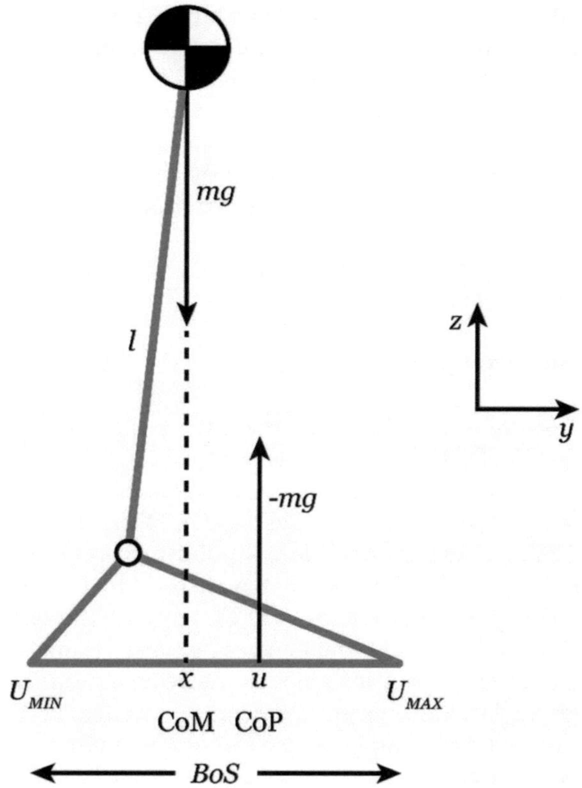

Figure 1.
Schematic of the inverted pendulum model [19]. Represented by a single mass on top of a stick, the body's center (CoM) is located at a distance l *from the ankle joint center. The center of pressure (CoP,* u*) identifies the location of the ground reaction force, which is placed relative to the vertical projection of the center of mass,* x*. the boundaries of the base of support (BOS), U_{MIN} and U_{MAX}, demonstrate the potential range of the CoP. The sagittal plane of the laboratory coordinate system,* y – z*, indicates that the line of walking progression is along the* y*-axis. Note this figure originated in [20].*

[21]. During gait, the COM naturally falls outside of the BOS. Dynamic measures of balance, such as the MOS, have been validated and are used often as part of gait analysis to describe both static and dynamic stability during ambulation [22]. The larger the MOS, the better the dynamic balance is within an individual. Maintaining a constant MOS may also reflect better dynamic control [23]. The medio-lateral (M/L) MOS increases as step width and single limb support time increase and is further dictated by the relationship between both limbs during the gait cycle [23–27].

Turning requires interlimb coordination, modification of movement patterns, coordination of posture and gait, and an increased level of executive function, all of which are impaired in persons with PD [8]. Falls during turns present a significant threat to the health and wellness of individuals with PD, adversely affecting an individual's independence and quality of life. Research regarding the control of dynamic balance during turns [20] may provide insight into where interventions may be targeted during therapeutic rehabilitation. Individuals with PD have impaired dynamic stability when completing turns while walking and may employ anticipatory adjustments in the gait and standing posture to maintain stability. Analysis of MOS variables in the gait cycle preceding a turn may provide information regarding biomechanical mechanisms underlying dynamic balance control in persons with PD. The purpose of this study was to analyze dynamic stability, as measured by the margin of stability, of the gait cycle just preceding a 90-degree turn during walking in persons with PD as compared to age-matched healthy adults. The results of this project may advance research into the biomechanical analysis of turning strategies and balance control in individuals with PD. This research may provide clinicians insight regarding balance strategies in persons with PD underlying their increased fall risk, as well as provide insight into possible fall prevention interventions.

2. Methods and materials

2.1 Recruitment and participants

This exploratory study was approved by the University Institutional Review Board (IRB reference#: 16–183-H). This project was part of a larger study that assessed a variety of tasks during walking that challenged dynamic balance in persons with PD. Since we collected data in a single session description of the data collection and reduction methods will be similar to previously published work [20].

Participants were recruited for this study from a local metropolitan area. Recruitment methods included flyers, presentations at PD support and exercise groups, speaking with medical professionals, and contacting participants from prior research projects. This study was advertised at local retirement communities, PD support groups, and through the local Parkinson's Association. An age- and gender-matched control group was formed by a sample of convenience including acquaintances, relatives, and friends of researchers.

Thirteen individuals with mild to moderate idiopathic PD and 10 controls (CON) were recruited for this study (**Figure 2**). Individuals at a high risk of falling and those with an indication of moderate cognitive impairment were excluded. The Berg Balance Scale was utilized to determine the risk of falling [28, 29]. A score of <36 and/or > 2 falls per month was considered a high fall risk. A score of <21 on the Montreal Cognitive Assessment was indicative of cognitive impairment among persons with PD [30]. Refer to the previous work by Alderink et al. [20] for a complete list of inclusion and exclusion criteria for this study. The decade of life, as stated in the criteria, was defined as 50–59, 60–69, etc.

Initial screening was conducted via phone interview. Upon meeting inclusion criteria, individuals were screened in person at the Biomechanics and Motor Performance Laboratory. Testing included completion of the Freezing of Gait Questionnaire (FOG-Q), Berg Balance Scale, and Montreal Cognitive Assessment. All participants were non-freezers for the FOG-Q, defined by responses to item 3, "Do you feel that

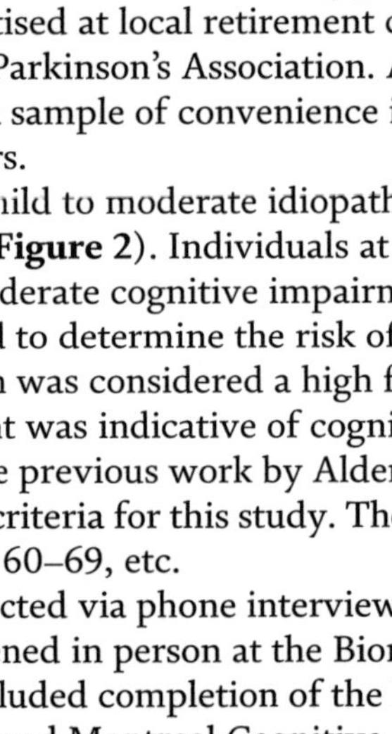

Figure 2.
Participant recruitment and participation flow diagram.

your feet get glued to the floor while walking, making a turn or when trying to initiate walking (freezing)?", which has been found to be a good screen for FOG frequency [31]. Additionally, the functional walking criteria were screened to ensure participants were able to complete the 300 m walk test unassisted. All participants provided written informed consent prior to inclusion in the study. Once consent was obtained, participants provided medical information including the history of their PD diagnosis, fall history, medical history, demographic information, and current medications. Participant demographic information is provided in **Table 1**.

2.2 Procedures

Walking and turning test procedures were completed during the "on" period of the participants' PD medication schedule as this reflects participants' usual functioning in a medicated state. Participants were asked to wear their normal walking shoes and clothing, including tight-fitting shorts, and sports bras for women, to enhance the exposure of anatomical landmarks and placement of motion capture markers. To prepare for optical motion capture trials, the following anthropomorphic measurements were taken: ankle, knee, elbow, and wrist width, hand thickness, leg length, and inter-ASIS (anterior-superior iliac spine) width; a shoulder offset was set at 3.5 cm for all participants. Lower extremity range of motion and strength assessment were completed using goniometric and myotome grading measures, respectively. Following the physical examination, 40 spherical markers (14 mm) were placed on landmarks on the body according to a modified version of the full-body Plug-in Gait (PIG) model (Appendix A). Modifications to the PIG model included additional markers on the heads of the 5th metatarsals and placement of the thigh markers on the mid-lateral thigh and tibial markers on the mid-lateral tibiae. Markers were placed by a single examiner and checked between walking sessions for any displacement. All examiners were trained by the lab director, but only a single examiner placed anatomical markers to eliminate inter-examiner variability [20].

Testing procedures included the following walking conditions 1) self-selected pace (which we defined, in this study, as Walking), 2) walking with 90-degree turns at a normal pace, 3) termination of gait and 4) walking with an obstacle crossing task. Dynamic balance variables were analyzed for the 90-degree turn and obstacle-crossing tasks in prior studies. This study focused its analysis of

	CON (n = 10)	PD (n = 11)*
Age (yrs.)	65.1 ± 7.4	67.5 ± 10.2
Gender (M:F)	9:1	10:1
Height (m)	1.8 ± 0.1	1.8 ± 0.1
Mass (kg)	86.4 ± 12.6	90.5 ± 17.5
BBS	55.3 ± 1.6	53.4 ± 3.3
MoCA	27.2 ± 2.4	25.9 ± 2.8
FOG-Q	Not Tested	2.7 ± 3.6

**Thirteen participated in this project but data from two participants did not meet the analysis criteria.*
BBS = Berg Balance Scale where the highest score of 56 indicates no balance deficits. MoCA = Montreal Cognitive Assessment where the highest score of 30 indicates no cognitive deficits. FOG-Q = Freezing Gait Questionnaire where the highest score of 24 indicates the most severe freezing of gait. Only the PD group was tested for FOG-Q.

Table 1.
Demographic means ± standard deviation for control (CON) and Parkinson's (PD) groups.

dynamic balance on gait cycles (referred to as Pre-turn walking) preceding the 90-degree turn.

Each participant completed a static standing calibration trial prior to beginning the walking trials to ensure correct marker positioning, create a subject-specific skeletal model, and identify where markers are in relation to joint centers or other bony landmarks. One walking trial was also completed and processed to examine joint kinematics prior to further data collection. Joint kinematics were graphed against a normative database to check for errors in the frontal and transverse knee and hip transverse plane kinematics that are indicative of marker placement error. If these kinematic data were abnormal, marker placement was revised and the static calibration and walking trial were repeated [20].

Participants practiced self-paced walking trials and were instructed to cleanly strike the force plates without targeting (i.e., looking at) them. All walking trials were completed during one session per participant. A gait belt was available for safety if needed per the examiner's discretion. Participants walked down a 10-meter walkway with three force plates aligned straight down the walkway. After sufficient practice, a

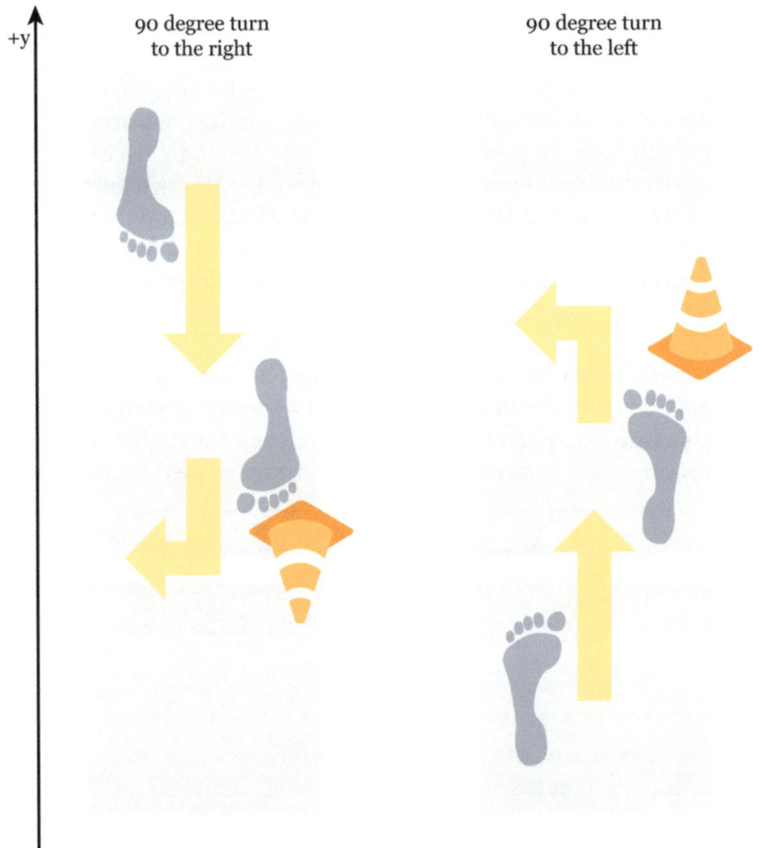

Figure 3.
Laboratory coordinate system, force plate set-up, and turning procedure (see [20]). For 90-degree turns to the right, participants walked in the negative y-direction and pushed off with their left foot in the negative x-direction. For a 90-degree turns to the left participants walked in the positive y-direction and pushed off with their right foot in the negative x-direction. The orange triangles represent the cones set up at the corner of the force plate.

minimum of five left and five right gait cycles with clean force plate strikes were collected [20].

For the 90-degree turn condition, a modified procedure based on Yang et al. was used [32]. Participants were allowed to look at the ground before the turn and target the force plate for this condition since we believed this simulated actual turning in normal ambulation of people with PD. To mark where the turn should occur, a cone was placed at the edge of the second force plate. Participants received standardized instructions for the trial. "For this task, we will be examining how you complete a 90-degree turn. Please walk at a comfortable speed along the walkway and turn to your (left/right) between the two pieces of tape placed next to the walkway. You will complete five trials of this task on each leg. I will demonstrate the task for you." For our study, we analyzed the kinematics at the force plate foot strike and gait of the step preceding the turn (**Figure 3**).

2.3 Instrumentation

Vicon Nexus v2.6.1 motion capture software was used to synchronize 16 T40S/MXF40/MX40 (120 Hz) cameras (Vicon Motion Systems LTD, Oxford, UK) and three AMTI (Advanced Mechanical Technology Inc., Watertown, MA) floor-embedded force plates (1200 Hz) and collect marker trajectories and ground reaction forces, respectively.

2.4 Data processing and reduction

Marker and ground reaction force data were post-processed and trials were trimmed to a single gait cycle in Vicon Nexus 2.6.1 (refer to [20] for details). Six representative gait cycles (three right and three left) were selected and exported to Visual3D (C-Motion, Inc., Germantown, MD) for the dynamic stability analysis. The MOS variables considered in this study were COM - COP inclination angle (**Figure 4**), COP - COM (**Figure 4**), COP - XCOM (**Figure 5**), and U_{MAX} - XCOM (**Figure 5**). The COM – COP inclination angle was defined relative to the laboratory coordinate system (**Figure 3**) as the spatial angle between the vertical projection of the COM and a line connecting the COM to the COP. The other variables were determined relative to the virtual foot coordinate system (**Figure 6**) in both the A/P and M/L directions.

The extrapolated center of mass (*XCOM*) was found using:

$$XCOM = COM + (V_{COM}/\omega_0) \tag{1}$$

where *COM* and V_{COM} are the instantaneous position and velocity of the total body *COM* and ω_0 is the angular eigenfrequency of the pendulum,

$$\omega_0 = \sqrt{g/l} \tag{2}$$

where g = 9.81 m/s^2 and l is the leg length (the distance from the *COM* to the ankle joint center in meters) [21].

Dynamic stability was calculated as the distance between the *XCOM* and the limits of the BOS:

$$MOS = U_{MAX} - XCOM \tag{3}$$

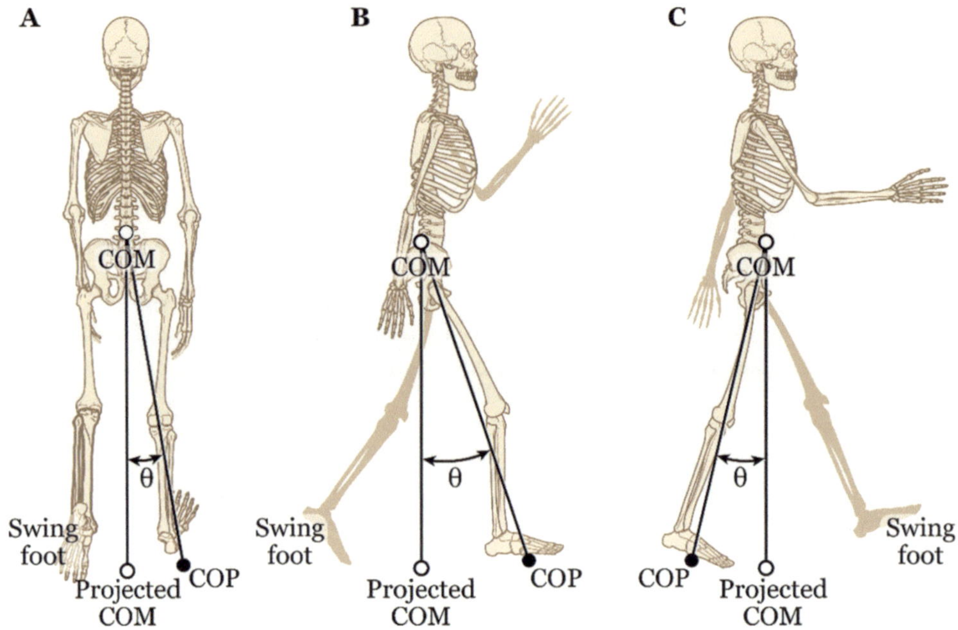

Figure 4.
Center of mass (COM) – Center of pressure (COP) inclination angle (θ) is the spatial angle between the vertical projection on the ground of the COM and a line connecting the COM to the most lateral position of the COP. A) Posterior view of the COM – COP inclination angle and the linear medio-lateral (M/L) distance between the vertical projection on the ground of the COM and location of the COP, i.e., COP – COM M/L, at midstance on the right B) lateral view of the COM – COP inclination angle and the linear antero-posterior (a/P) distance between the projection on the ground of the COM and location of the COP, i.e., COP – COM a/P, at first double-support, and C) lateral view of the COM – COP inclination angle and COP – COM a/P at second double-support (see [20]).

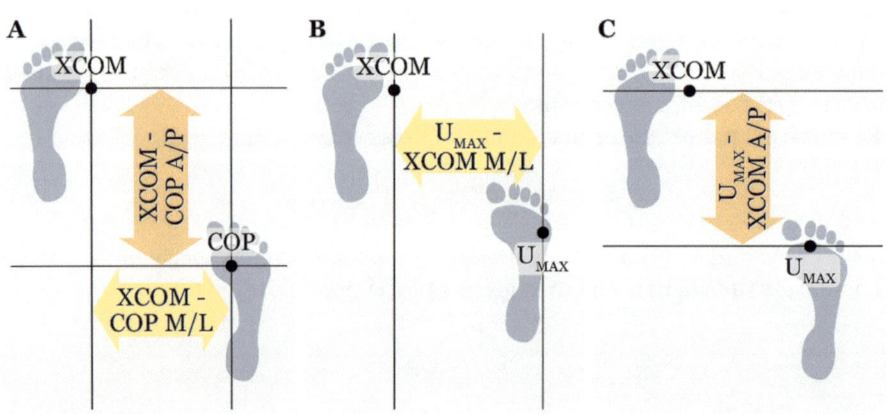

Figure 5.
Schematic illustrating the determination of the linear distance between the extrapolated center of mass (XCOM) and the center of pressure (COP), or U_{MAX}. A) the linear distance between the COP and XCOM in medio-lateral (M/L) and antero-posterior (a/P) directions, i.e., COP – XCOM M/L and COP - XCOM a/P, respectively, B) the linear distance between U_{MAX} (i.e., head of the 5th metatarsal) and XCOM in the medio-lateral direction, i.e., U_{MAX} – XCOM M/L, and C) the linear distance between U_{MAX} (i.e., between the heads of the 1st and 2nd metatarsals) and XCOM in the antero-posterior direction, i.e., U_{MAX} – XCOM a/P (see [20]).

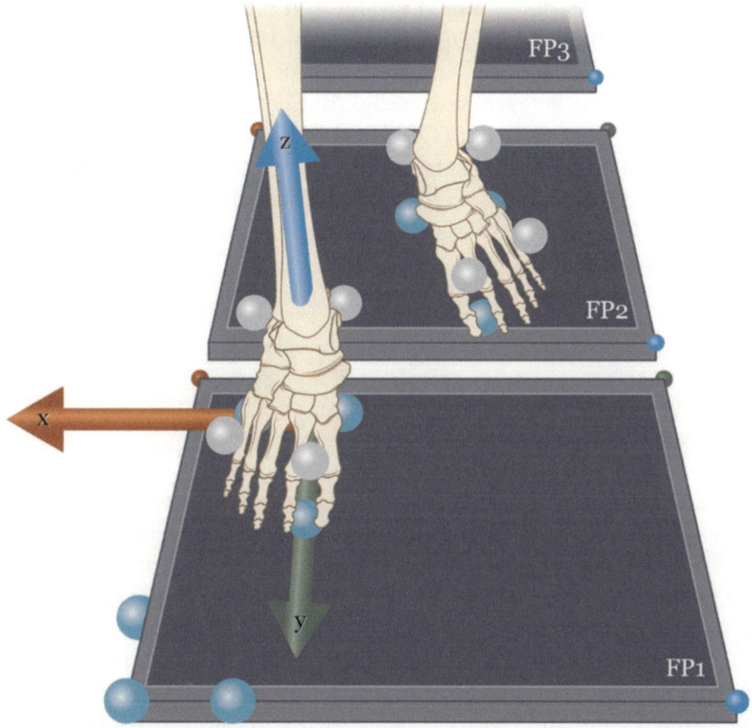

Figure 6.
A virtual foot coordinate system (CS) is defined by markers (light blue) over the medial and lateral malleoli and the toe (between the first and second metatarsal heads). The markers were projected onto the floor (dark blue) of the lab to align the CS parallel to the floor for estimation of the position of the center of pressure relative to the foot. The origin of the virtual foot CS is located at the ankle joint center (midway between the medial and lateral malleoli) projected to the floor. For the right foot, the right-handed CS is positive-x (red, medio-lateral), positive-y (green, antero-posterior), and positive-z (blue, superior-inferior). FP1 = force plate one; FP2 = force plate two; FP3 = force plate three.

where U_{MAX} in the anterior direction is defined by the toe marker and U_{MAX} in the lateral direction is defined by the 5th metatarsal marker.

The MOS variables were analyzed at three points in the gait cycle: first double-support (FDS) at 6% of stance, midstance (MS) at 50% of stance, and second double-support (SDS) at 94% of stance [20]. Spatiotemporal parameters (ST) of velocity, stride width, stride length, step length, cadence, stance time, and swing time were also calculated. All study variables were exported to spreadsheets and R (version 4.2.2 (2022-2110-31)) [33] and RStudio (Version 2023.03.0 + 386) [34] statistical software for descriptive and statistical analyses.

2.5 Statistical analysis

Preliminary statistical analysis demonstrated no significant differences in dynamic balance measures between the right and left limbs, therefore, we pooled the right and left gait cycle data. Statistical analyses for primary and secondary variables were performed using nlme [35] and emmeans [36] packages in R/RStudio. MOS variables were analyzed with a three-level mixed model ANOVA. The fixed factors were: phase (FDS, MS, and SDS), condition (CON and PD), and their interaction. Phase was modeled as a repeated measure within gait cycle and subject. Spatiotemporal variables were analyzed with a two-level mixed model ANOVA. The fixed factors were: task (Pre-Turn and Walking), condition (CON and PD), and their interaction. Both models used subject as a random intercept and the gait cycle as the observational unit. Primary and secondary variables were also analyzed with post hoc analyses using two-sample t-tests ($\alpha = 0.05$). The effect size was reported as the estimated difference between the means (Appendix B).

3. Results

3.1 Spatiotemporal gait parameters

Since we suspected that participants with PD may have altered their gait when anticipating a turn, we analyzed ST parameters between conditions (i.e., between PD and CON groups), as well as between tasks: Pre-turn, i.e., defined as the gait cycle immediately preceding the turn, and Walking, i.e., defined as a gait cycle during self-selected normal walking. When comparing Pre-turn to Walking within the CON group, all variables were significantly different (**Table 2**). Likewise, within the PD group, all ST parameters were significantly different between the two tasks ($p < 0.05$) (**Table 3**). For the Pre-turn task, there were no significant differences between CON and PD for any of the ST parameters (**Table 4**). All ST parameters were significant for task, demonstrating that these parameters were different for Pre-turn and Walking (**Table 5**); however, there were no differences for condition. Double limb

	Pre-Turn	Walking	P-Value*	Effect size†	95% CI
Double limb support (s)	0.355 (0.01526)	0.299 (0.01526)	<0.001	0.0564	0.04640, 0.0665
Cycle time (s)	1.209 (0.02943)	1.133 (0.02958)	<0.001	0.0758	0.06232, 0.0893
Stance time (s)	0.774 (0.02154)	0.723 (0.02169)	<0.001	0.0513	0.04009, 0.0626
Swing time (s)	0.430 (0.00811)	0.417 (0.00828)	<0.001	0.0130	0.00612, 0.0199
Step length (m)	0.662 (0.02089)	0.711 (0.02108)	<0.001	−0.0490**	−0.06143, −0.0366
Stride length (m)	1.337 (0.04536)	1.434 (0.04553)	<0.001	−0.0965**	−0.11391, −0.0791
Cadence (steps/min)	100.680 (2.51985)	105.605 (2.54318)	<0.001	−4.9250**	−6.46580, −3.3841
Velocity (m/s)	1.122 (0.03215)	1.256 (0.03215)	<0.001	−0.1335**	−0.15141, −0.1156

P-values are for a t-test of the difference in means between Pre-Turn and Walking for the control group. Likewise, we are reporting the 95% confidence interval (CI) for the difference between the means. Negative effect size indicates the Pre-turn value was less than the Walking value.†The effect size was determined relative to the differences between the means for the Pre-Turn and Walking tasks.*

Table 2.
Mean (standard deviation) for spatiotemporal (ST) gait parameters for the control group with task (1) pre-turn and (2) walking*.*

support, swing time, cadence, and velocity were significant for task and condition, representing a difference between PD and CON within Pre-turn and Walking (**Table 5**).

3.2 Measures of dynamic stability: a qualitative description of the shapes of the curves for the metrics of dynamic balance

Visual inspection of the graphical representations for the dynamic balance metrics revealed similar trends in the direction of change of the curves, but some differences in the magnitude of changes during the stance phase of gait for both the Parkinson's

	Pre-Turn	Walking	P-Value*	Effect size†	95% CI
Double limb support (s)	0.358 (0.01477)	0.326 (0.01477)	<0.001	0.0324	0.0252, 0.0396
Cycle time (s)	1.216 (0.02952)	1.133 (0.02961)	<0.001	0.0828	0.0689, 0.0968
Stance time (s)	0.780 (0.02160)	0.730 (0.02167)	<0.001	0.0498	0.0392, 0.0603
Swing time (s)	0.437 (0.00863)	0.401 (0.00871)	<0.001	0.0360	0.0286, 0.0434
Step length (m)	0.663 (0.02089)	0.707 (0.02098)	<0.001	−0.0448**	−0.0557, −0.0339
Stride length (m)	1.338 (0.04533)	1.421 (0.04540)	<0.001	−0.0834**	−0.0987, −0.0682
Cadence (steps/min)	98.878 (2.53787)	106.755 (2.55478)	<0.001	−7.8766**	−9.5368, −6.2164
Velocity (m/s)	1.100 (0.03212)	1.266 (0.03209)	<0.001	−0.1662**	−0.1824, −0.1500

P-values are for a t-test of the difference in means between Pre-Turn and Walking tasks. Likewise, we are reporting the 95% confidence interval (CI) for the difference between the means. Negative effect size indicates the Pre-turn value was less than the Walking value.†The effect size was determined relative to the differences between the means for the Pre-Turn and Walking groups.*

Table 3.
Mean (standard deviation) spatiotemporal (ST) gait parameters for the Parkinson's group with task (1) pre-turn and (2) walking.

	CON	PD	P-Value*	Effect size†	95% CI
Double limb support (s)	0.355 (0.01526)	0.358 (0.01477)	0.892	−0.002921**	−0.0480, 0.0421
Cycle time (s)	1.209 (0.02943)	1.216 (0.02952)	0.869	−0.006984**	−0.0953, 0.0814
Stance time (s)	0.774 (0.02154)	0.780 (0.02160)	0.86	−0.005486**	−0.0702, 0.0592
Swing time (s)	0.430 (0.00811)	0.437 (0.00863)	0.593	−0.006455**	−0.0316, 0.0186
Step length (m)	0.662 (0.02089)	0.663 (0.02089)	0.986	−0.000537**	−0.0632, 0.0621
Stride length (m)	1.337 (0.04536)	1.338 (0.04533)	0.992	−0.000654**	−0.1366, 0.1353
Cadence (steps/min)	100.680 (2.51985)	98.878 (2.53787)	0.621	1.801392	−5.7802, 9.3830
Velocity (m/s)	1.122 (0.03215)	1.100 (0.03212)	0.633	0.022151	−0.0742, 0.1185

P-values are for a t-test of the difference in means between CON and PD groups. Likewise, we report the 95% confidence interval for the difference between the means. Negative effect size indicates the CON value was less than the PD value.†The effect size was determined relative to the differences between the means for the CON and PD groups.*

Table 4.
Mean (standard deviation) spatiotemporal (ST) gait parameters for pre-turn for the control (CON) and Parkinson's (PD) groups.

	Task		Condition		Task: Condition	
	F	P-value	F	P-value	F	P-value
Double limb support (s)	186.3	<0.001	0.491480	0.493	14.6754	<0.001
Cycle time (s)	258.9	<0.001	0.000351	0.985	0.5077	0.477
Stance time (s)	167.1	<0.001	0.018181	0.894	0.0396	0.842
Swing time (s)	86.3	<0.001	0.155876	0.698	20.0854	<0.001
Step length (m)	125.9	<0.001	0.000246	0.988	0.2498	0.618
Stride length (m)	235.0	<0.001	0.000362	0.985	1.2405	0.266
Cadence (steps/min)	120.4	<0.001	0.008546	0.927	6.5891	0.0109
Velocity (m/s)	618.6	<0.001	0.007701	0.931	7.1445	0.00816

Table 5.
Fixed effects ANOVA of condition (Parkinson's, control), task (pre-turn, walking), and interaction for spatiotemporal (ST) gait parameters.

and control groups (**Figures 7** and **8**). In general, it's apparent that there was greater trial variation in COP – COM, COP - XCOM, and U_{MAX} – XCOM metrics in the medio-lateral direction for both PD and CON groups. To interpret these figures and to interpret the data presented in subsequent sections, the following guidelines will help:

- A larger magnitude of change in the distance between the center of mass and the base of support is consistent with a larger margin of stability.
- For the COP – COM A/P, COP – XCOM A/P, and U_{MAX} – XCOM A/P metrics a positive value indicates that the COP (and U_{MAX}) is anterior to the COM; conversely, a negative value indicates that the COP lies posterior to the COM. Thus, as participants moved from FDS through MS to SDS, the COP progressed from in front of (anterior) the COM to behind (posterior).
- For the COP – COM M/L, COP – XCOM M/L, and U_{MAX} – XCOM M/L metrics a positive value indicates that the COP (and U_{MAX}) is lateral to the COM, i.e., the COM lies within the base of support, whereas a negative value indicates that the COP lies medial to the COM.
- For the COP – COM inclination angle increasing positive values are consistent with a greater angle, thus a greater distance between the COP and COM. For example, as expected, the inclination angle was greater at first and second double support.

3.3 Metrics of dynamic balance

3.3.1 COP – COM antero-posterior and medio-lateral

There were no significant differences in the COP – COM metric between the PD and CON groups at FDS, MS, or SDS in the antero-posterior (A/P) direction (**Table 6**). However, there was an interesting result in the medio-lateral (M/L) direction at MS, with the PD (20.24 ± 9.880 mm) metric greater than the CON

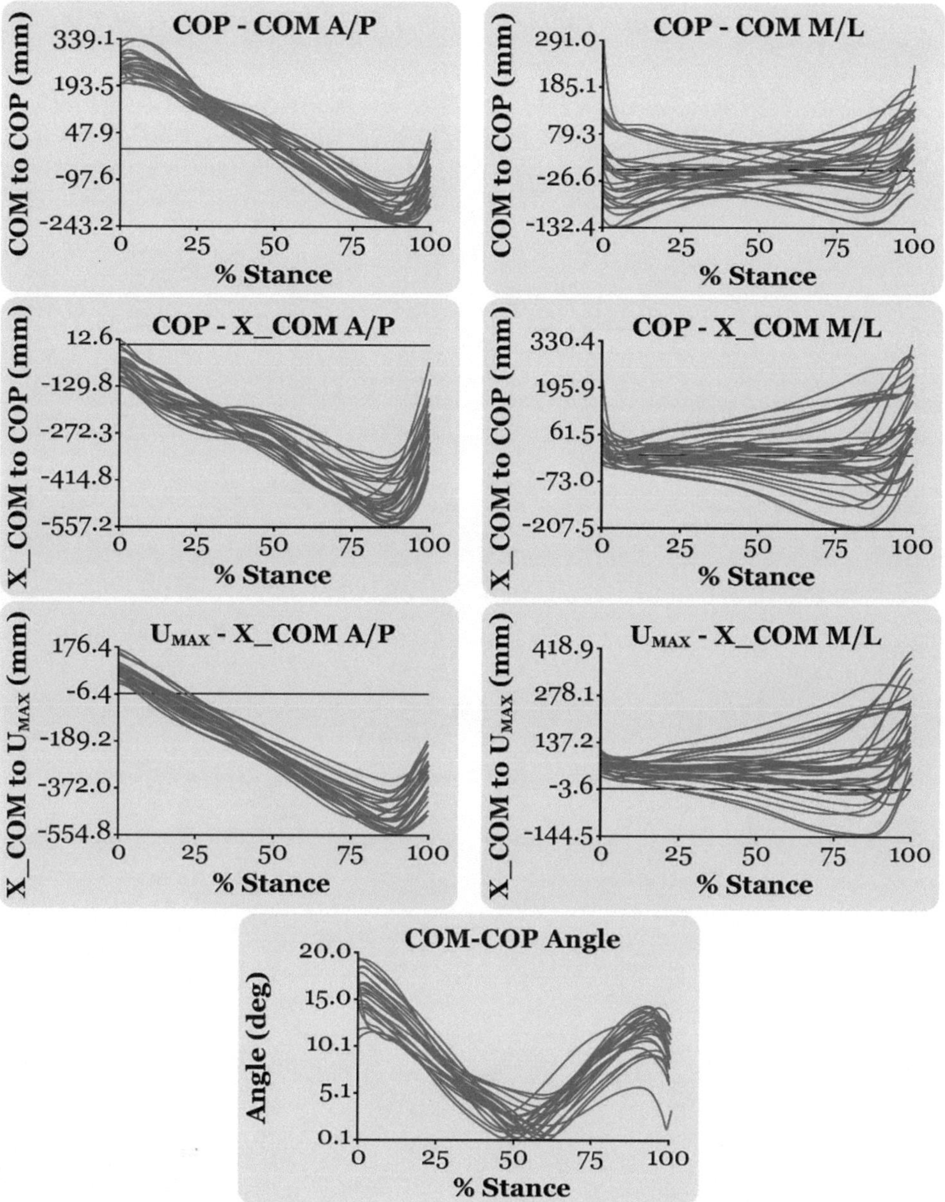

Figure 7.
Dynamic balance metrics for the right limb for all subjects during percent stance of the gait cycle for the a) CON and B) PD groups. For the a/P COP - COM, positive values indicate that the COP is anterior to the COM. In M/L COP - COM, increasing positive values indicate that the COP is more lateral to the COM. The same is true for the remaining pairs, with positive COP and positive U_{MAX} indicating that the toe marker is anterior to XCOM in the a/P direction, and increasing positive values indicating that the COP and U_{MAX} are lateral to XCOM in the M/L direction.

(−1.30 ± 10.128 mm) metric (**Table 7**). The negative value for the CON group suggests that the COM was outside the base of support, i.e., the COP, that is, the CON group's COM ventured beyond the BOS yet balance was maintained. Thus, the larger value for COP - COM at MS in PD may reflect a more stable strategy to keep the COM

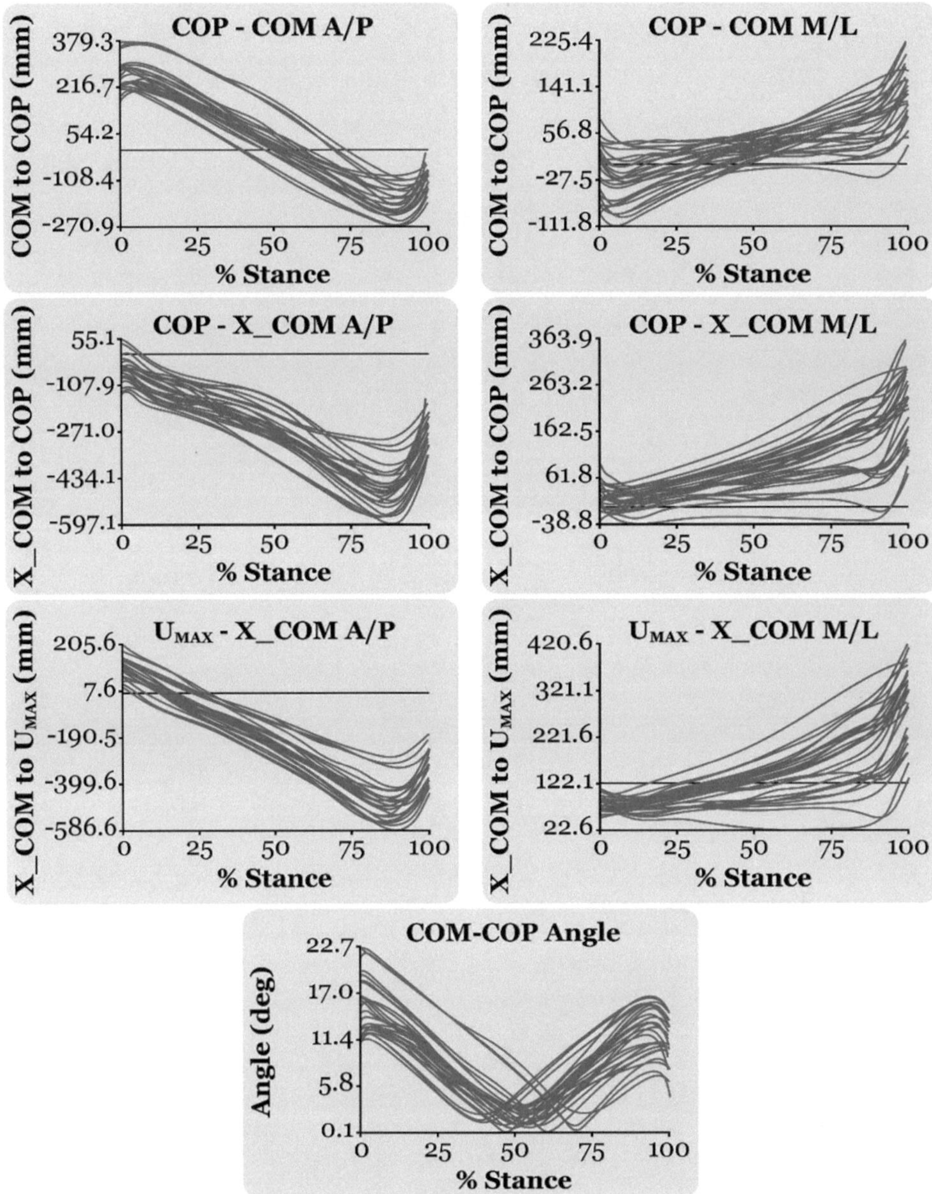

Figure 8.
Dynamic balance metrics for the right limb for all subjects during percent stance of the gait cycle for the a) CON and B) PD groups. For the a/P COP-COM, positive values indicate that the COP is anterior to the COM. In M/L COP - COM, increasing positive values indicate that the COP is more lateral to the COM. The same is true for the remaining pairs, with positive COP and positive U_{MAX} indicating that the toe marker is anterior to XCOM in the a/P direction, and increasing positive values indicating that the COP and U_{MAX} are lateral to XCOM in the M/L direction.

within the BOS. There was also a significant difference in the COP - COM M/L variable at SDS between the PD (86.52 ± 9.880 mm) and CON (35.37 ± 10.128 mm) groups ($p = 0.00198$). The larger value for the PD group at both MS and SDS may be a compensatory change to increase dynamic stability and balance.

Significant differences were found for COP – COM in both the A/P and M/L directions for the main effect of condition ($p < 0.001$), suggesting a consistent difference in means between the PD and CON groups across phases. However, differences between groups were only seen in the M/L direction for phase:condition interaction ($p = 0.001$) (**Table 8**). Thus, in the M/L direction, the COP - COM distances were significantly different between the CON and PD groups. Furthermore, these differences were apparent at MS and SDS between the CON and PD groups (**Figure 9**). The most notable difference between groups appeared to be at SDS (**Figure 9**). The PD group had a larger median value for COP - COM M/L at SDS compared to the CON group. Since SDS occurs just prior to the turn this may be a more unstable time point in the gait cycle for the PD group and increases in MOS may assist in maintaining balance.

3.3.2 COP – XCOM antero-posterior and medio-lateral

There were no significant differences between the PD and CON groups at FDS, MS, and SDS time points for COP - XCOM in the A/P direction (**Table 7**) or in the M/L direction at FDS (**Table 8**). However, there were significant differences between PD and CON for the COP - XCOM metric in the M/L direction at MS and SDS (**Table 8**). At MS, the mean COP - XCOM was significantly larger for the PD (50.50 ± 13.527 mm) as compared to the CON group (2.91 ± 14.152 mm) ($p = 0.0257$). At SDS, the mean COP - XCOM was significantly larger for the PD (140.98 ± 13.527 mm) than the CON group (29.75 ± 14.152 mm) ($p < 0.001$). The

	PD	CON	P-Value*	Effect size	95% CI
FDS					
COP - COM	258.51 (6.785)	254.86 (6.975)	0.712	−3.655	−24.098, 16.789
COP - XCOM	−93.81 (9.060)	−109.53 (9.210)	0.239	−15.726**	−42.868, 11.416
U_{Max} - XCOM	85.48 (9.974)	58.72 (10.186)	0.0768	−26.757**	−56.708, 3.194
MS					
COP - COM	19.76 (6.785)	23.17 (6.975)	0.73	3.405	−17.038, 23.849
COP - XCOM	−265.37 (9.060)	−290.64 (9.210)	0.0661	−25.272**	−52.414, 1.870
U_{Max} - XCOM	−202.89 (9.974)	−226.29 (10.186)	0.118	−23.397**	−53.348, 6.553
SDS					
COP - COM	−193.00 (6.785)	−195.30 (6.975)	0.816	−2.293**	−22.736, 18.150
COP - XCOM	−422.51 (9.060)	−424.60 (9.210)	0.873	−2.094**	−29.236, 25.048
U_{Max} - XCOM	−423.11 (9.974)	−421.09 (10.186)	0.889	2.020	−27.931, 31.971

**CON and PD A/P dynamic balance variables were determined for all participants walking at their self-selected pace before they initiated a 90-degree turn during first double support (FDS), mid-stance (MS), and second double support (SDS). P-values are for a t-test of the difference in means between CON and PD groups. The effect size was determined relative to the differences between the means for the CON and PD groups. Likewise, we are reporting the 95% confidence interval for the difference between the means. **Negative effect sizes indicate that the CON group values were less than the PD group values.*

Table 6.
Mean (standard deviation) expressed in millimeters for Antero-posterior dynamic balance variables for control (CON) and Parkinson's (PD) groups during the pre-turn gait cycle.

	PD	CON	P-Value*	Effect size	95% CI
FDS					
COP - COM	−25.64 (9.880)	−27.51 (10.128)	0.897	−1.865**	−31.591, 27.862
COP - XCOM	14.57 (13.527)	9.67 (14.152)	0.805	−4.903**	−46.033, 36.227
U_{Max} - XCOM	74.96 (12.781)	62.50 (13.524)	0.511	−12.464**	−51.558, 26.629
MS					
COP - COM	20.24 (9.880)	−1.30 (10.128)	0.145	−21.538**	−51.265, 8.188
COP - XCOM	50.50 (13.527)	2.91 (14.152)	0.0257	−47.586**	−88.716, −6.456
U_{Max} - XCOM	103.83 (12.781)	55.16 (13.524)	0.0175	−48.667**	−87.761, −9.574
SDS					
COP - COM	86.52 (9.880)	35.37 (10.128)	0.00198	−51.144**	−80.870, −21.417
COP - XCOM	140.98 (13.527)	29.75 (14.152)	<0.001	−111.233**	−152.363, −70.103
U_{Max} - XCOM	204.23 (12.781)	93.49 (13.524)	<0.001	−110.746**	−149.840, −71.653

**CON and PD M/L dynamic balance variables were determined for all participants walking at their self-selected pace before they initiated a 90-degree turn during first double support (FDS), mid-stance (MS), and second double support (SDS). P-values are for a t-test of the difference in means between CON and PD groups. The effect size was determined relative to the differences between the means for the CON and PD groups. Likewise, we are reporting the 95% confidence interval for the difference between the means. **Negative effect sizes indicate that the CON group values were less than the PD group values.*

Table 7.
Mean (standard deviation) expressed in millimeters for Medio-lateral dynamic balance variables for control (CON) and Parkinson's (PD) groups during the pre-turn gait cycle.

		Phase**		Condition*		Phase: Condition	
		F	P-value	F	P-value	F	P-value
COP - COM	A/P	0.0088	0.926	10,907	<0.001	0.712	0.492
	M/L	3.4724	0.0788	267	<0.001	18.328	<0.001
COP - XCOM	A/P	1.3708	0.257	4317	<0.001	5.520	0.0046
	M/L	8.5802	0.0090	187	<0.001	52.804	<0.001
U_{Max} - XCOM	A/P	1.3728	0.257	11,094	<0.001	10.502	<0.001
	M/L	10.673	0.0042	223	<0.001	42.560	<0.001

**CON and Parkinsons M/L dynamic balance variables were determined for all participants walking at their self-selected pace before they initiated a 90-degree turn.*
*** Phase is used to categorize first double support (FDS), mid-stance (MS), and second double support (SDS).*

Table 8.
Fixed effects ANOVA of condition (PD, CON) and phase (FDS, MS, SDS) for anterior-posterior (a/P) and medio-lateral (M/L) dynamic balance variables during the pre-turn gait cycle.

smaller value for the CON group suggests that XCOM was closer to the COP at both MS and SDS. For the PD group at MS and SDS, XCOM was further from the BOS, which may suggest compensatory changes to increase MOS as a strategy to maintain balance prior to a turn.

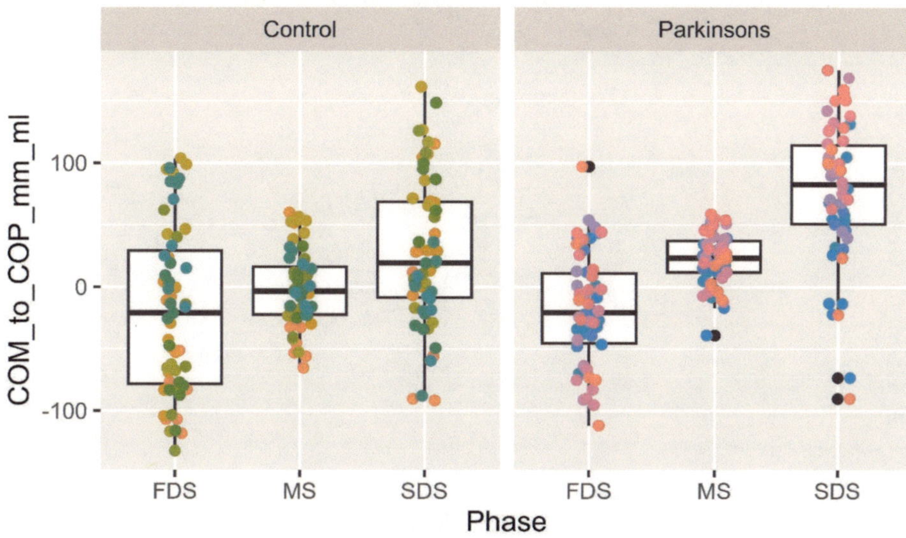

Figure 9.
Boxplots with overlayed data points illustrating subject by color, condition (control and Parkinsons), phase (first double support (FDS), midstance (MS), and second double support (SDS)), and interaction of the two. Median values, reported in millimeters, were significantly different in the M/L direction for COP to COM.

There were significant differences due to condition ($p < 0.001$) for COP – XCOM in the A/P and M/L directions (**Table 9**). Furthermore, the COP – XCOM metric showed differences in both the A/P ($p = 0.0046$) and M/L ($p < 0.001$) directions for the interaction phase:condition. Larger values for the PD group, particularly at MS and SDS (**Figure 10**) suggest that XCOM was further from the BOS. The most notable difference in phase and condition interaction appeared to be at SDS (**Figure 10**). For the PD group, increasing MOS as the gait cycle progressed from FDS to SDS may represent motor planning changes in preparation for a turn.

3.3.3 U_{MAX} – XCOM antero-posterior and medio-lateral

As with the previous two metrics, there were no significant differences in U_{Max} – XCOM between the PD and CON groups in the A/P direction (**Table 6**).

	PD*	CON*	P-Value	Effect size**	95% CI
FDS	14.28 (0.387)	14.59 (0.393)	0.572	0.318	−0.841, 1.477
MS	2.25 (0.387)	2.40 (0.393)	0.783	0.154	−1.005, 1.313
SDS	13.04 (0.387)	12.02 (0.393)	0.0817	−1.017**	−2.176, 0.142

**CON and PD groups' COP - COM inclination angles were determined for all participants walking at their self-selected pace before they initiated a 90-degree turn during first double support (FDS), mid-stance (MS), and second double support (SDS). P-values are for a t-test of the difference in means between CON and PD groups, the effect size was determined relative to the differences between the means for both groups. Likewise, we are reporting the 95% confidence interval for the difference between the means.*
***Negative effect sizes indicate that the CON group values were less than the PD group values.*

Table 9.
Mean (standard deviation) expressed in degrees for COP - COM inclination angle for FDS, MS, and SDS for control (CON) and Parkinson's (PD) groups during the pre-turn gait cycle.

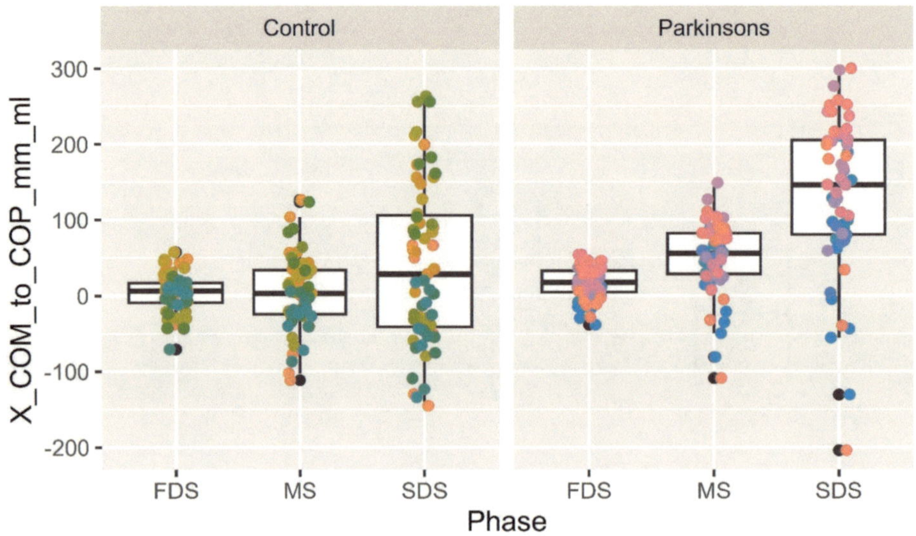

Figure 10.
Boxplots with overlayed data points illustrating subject by color, condition (control and Parkinson's), phase (first double support (FDS), midstance (MS), and second double support (SDS)), and interaction of the two. Median values, reported in millimeters, were significantly different in the M/L direction for XCOM to COP.

However, at FDS in the A/P direction, the positive values for both PD and CON indicate that the toe marker, or U_{MAX}, was anterior to the XCOM meaning that the XCOM was within the base of support (BOS). Yet the margin for PD was larger with an effect size of approximately 2.5 cm (i.e., 26.757 mm). This large effect size and nearly significant result (p = 0.0768) suggest a potential clinically meaningful difference.

The U_{MAX} - XCOM variable in the M/L direction showed significant differences between PD and CON at MS and SDS (**Table 8**). At MS the U_{MAX} - XCOM was significantly greater in the PD group (103.83 ± 12.781 mm)) compared to the CON group (55.16 ± 13.524 mm (p = 0.0175). Likewise, the U_{Max} – XCOM SDS for the PD (204.23 ± 12.781) group was greater than for CON (93.49 ± 13.524). The smaller values for the CON group indicate that XCOM was closer to the U_{MAX} at MS and SDS. This suggests that XCOM was further from the BOS in the PD group compared to CON.

For condition and phase:condition there were significant differences in the U_{Max} – XCOM metric in the A/P and M/L directions ($p < 0.001$). However, only M/L directional differences were noted in phase (**Table 8**).

3.3.4 COM – COP inclination angle

The COM - COP inclination angle was not significantly different between CON and PD groups at FDS, MS, and SDS (**Table 9**). Findings did not reveal a significant difference for phase, but did note differences for condition ($p < 0.001$) and phase: condition (p = 0.00129) (**Table 10**). The COM - COP inclination angle was larger at FDS and SDS due to the increased distance between the right and left limbs at these points during the stance phase of gait, and smaller at MS as the two limbs approached each other for both PD and CON.

3.3.5 Relationship between the metrics of dynamic balance

COP - COM, COP - XCOM, and U_{MAX} - XCOM in A/P and M/L directions were the variables we used to measure MOS in this study. It is notable that the trajectories and slopes for these variables from one representative participant from the CON group were similar (**Figure 11A**). Overlaid curves from one representative participant from the PD group show similar trends (**Figure 11B**). All three variables, COP - COM, COP - XCOM, and U_{MAX} - XCOM, are methods to describe dynamic balance and MOS. The extrapolated COM (XCOM) accounts for the horizontal velocity of the COM during walking. The horizontal velocity component that is considered with XCOM and not COM might be one explanation for differences between COP - COM and the other MOS variables. For example, it appears that consideration of velocity appears to decrease the magnitude of the MOS (**Figure 11**). Thus, the larger difference between the COP - COM curve and U_{MAX} - XCOM or COP - XCOM curves could be due to the consideration of horizontal velocity, while the curves for U_{MAX} - XCOM and COP - XCOM were more similar. The margin of stability variables using XCOM demonstrated a larger change from FDS to SDS. The magnitude of change for each curve appears to be greater within the A/P direction than in the M/L direction. In the M/L direction, COP - COM, COP - XCOM, and U_{MAX} - XCOM were more similar with less differentiation with the use of XCOM compared to the A/P direction.

4. Discussion

Balance challenges, such as performing a sharp turn while walking can lead to falls and potentially debilitating injuries in persons with Parkinson's disease. In an effort to better understand fall-related mechanics, this study investigated select biomechanical factors related to the control of dynamic stability during walking and turning. The purpose of this study was to examine the dynamic stability, as measured by the margin of stability (MOS), and other related biomechanical metrics of balance stability, during the gait cycle just preceding the 90-degree turn in persons with PD as compared to healthy age-matched controls. In general, during Pre-turn gait cycles, the PD group demonstrated significantly greater M/L MOS metrics at MS and SDS.

The inverted pendulum model has informed human balance research since its introduction to the field [19, 37]. The relationship between the center of mass and related extrapolated center of mass, and center of pressure, has been investigated under several different scenarios. The metrics derived from these relationships have been shown to be valid and reliable measures for the study of the balance of persons

	Phase**		Condition*		Phase: Condition	
	F	P-value	F	P-value	F	P-value
COM - COP Inclination Angle	0.13061	0.722	2223	<0.001	6.847	0.00129

**Parkinson's and Control COM - COP inclination angles were determined for all participants walking at their self-selected pace before they initiated a 90-degree turn.*
***Phase is used to categorize first double support (FDS), mid-stance (MS), and second double support (SDS).*

Table 10.
Fixed effects ANOVA of condition (Parkinson's, control) and phase (FDS, MS, SDS) for COM - COP inclination angle during pre-turn gait cycle.

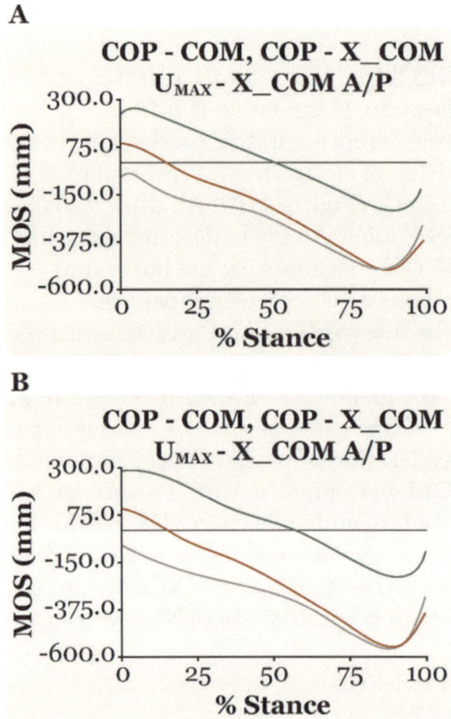

Figure 11.
Overlaid dynamic balance variables of a representative participant from a) control and B) Parkinson's groups for a/P and M/L COP - COM, COP - XCOM, U_{MAX} - XCOM. Green line = COP - COM, black line = COP - XCOM, red line = U_{MAX} - XCOM. Percent (%) stance of the gait cycle accounts for FDS at 6.0%, MS at 50.0%, and SDS at 94.0%.

with neurologic disorders that affect gait [21]. Hof suggested that maintaining a constant MOS resulted in a more stable gait [24]. Buurke et al. found that interlimb coordination played a greater role in mediolateral dynamic stability, particularly in patients with balance deficits [23]. Although both intrinsic and extrinsic factors affect dynamic balance, increasing SLS time and changes in the BOS and control of the MOS are strategies used to increase stability during walking.

Since there is a complex relationship between the spatiotemporal parameters of gait and dynamic balance, it is likely that the metrics of balance during walking will change if there are changes in the ST parameters. To our knowledge, this study is the first to compare ST gait parameters of normal self-paced walking (here described as Walking) to the gait cycle preceding a turn (here described as Pre-turn). All of the ST variables examined were significantly different between Walking and Pre-turn for both the CON and PD groups. These findings may reflect that preparatory alterations in gait and balance control in anticipation of making a turn were evident both in persons with PD and healthy age-matched controls.

When comparing ST variables between CON and PD exclusively during Pre-turn, the variables were not significantly different from each other. This may indicate that in individuals with mild to moderate PD, changes in the ST parameters during turns may be more subtle or minor. Double limb support time approached significance between the PD and CON groups, as the PD group demonstrated longer double limb support times than healthy controls. A systematic review with meta-analysis of gait changes in persons with PD reported that increased double limb support time is a common result seen in studies of PD gait [38]. Corroboration of our findings and those of Zanardi et al. suggest that an increased double limb support time in persons with Parkinson's disease may be an adaptation to improve stability by reducing single-limb support time.

When considering the effects of task: condition during Pre-turn walking (**Table 6**) there were significant differences between groups in double limb support and swing time, as well as cadence and velocity. These spatiotemporal differences between persons with PD and healthy controls may represent early motor changes in gait found with PD. For example, decreased swing time, and associated increases in double limb support time, may be an early symptom of an emerging hypometric gait pattern that is a hallmark sign of Parkinson's disease. Furthermore, decreases in cycle time and cadence may be associated with an eventual decrease in gait velocity, known as bradykinesia in persons with PD [3, 39]. King et al. suggested that the neural pathways relating to bradykinesia may affect even those with very mild PD due to the highly complex nature and executive function requirements of turning while walking [16].

We anticipated that there might be differences in the A/P and M/L MOS metrics between individuals with PD and healthy controls during the gait cycle preceding a turn. To our knowledge, this is the first study to demonstrate selected differences in these metrics of dynamic balance. For the COP - COM metric in the A/P direction, there were no differences between the CON and PD groups. In contrast, there were notable differences in COP - COM in the M/L direction between the two groups. The COP-COM M/L metric at MS for the CON group had a negative value, indicating that the COM was outside the BOS, i.e., COP. This finding suggests a tendency for a less stable position. In contrast, at SDS the PD group's COM was further from the BOS reflective of a more stable position. This finding may suggest that the healthy controls were able to maintain dynamic balance with a smaller M/L MOS compared to the PD group, or that the PD group increased their M/L MOS to adapt to changes in dynamic balance that may predispose them to a fall. The control group's COM-COP distance was smaller, and their COM ventured beyond the COP. These findings may reflect normal and robust balance abilities, which indicates that healthy individuals can control their dynamic balance even when the COM moves beyond the BOS during this walking task demand.

For the COP - XCOM metric, there were no differences between groups in the A/P direction; however, there were significant differences between groups in the M/L direction at both MS and SDS. At these points in the stance phase, the PD group's XCOM was further from the BOS suggesting a more stable position. The PD group may have made different adaptations to preserve their balance than healthy controls in preparation for a turn. In contrast to our findings, the study by Mellone et al. found that participants with mild to moderate PD had a narrower BOS than age-matched healthy adults, resulting in a smaller mean distance between XCOM and BOS. This would indicate a level of instability [8]. These ideas may suggest that individuals with PD could increase their BOS as a compensatory strategy to maintain balance and therefore, increase their MOS. Practitioners might use these results to inform clinical practice by demonstrating the importance of a wider BOS during turns and ambulatory tasks.

In the medio-lateral direction, the U_{MAX} - XCOM was significantly larger in PD compared to CON at MS and SDS. The PD group's XCOM was further from the BOS, once again suggesting a more stable position. It has been proposed that in persons with stroke-related balance impairments, individuals are more likely to achieve balance following a perturbation by changing their COM position rather than their COM velocity (XCOM) compared to their BOS [40]. Since XCOM is velocity-dependent, it is possible that an interaction between COM and velocity could reduce the U_{MAX} - XCOM metric. Buurke et al. suggested that individuals with impaired dynamic balance can improve M/L MOS by improving inter-limb coordination [23]. PD-related gait changes that adversely affect inter-limb coordination are another possible explanation for differences found in MOS. Further research on this metric of dynamic balance in persons with PD-related gait impairments is warranted.

There were no differences found in the COM-COP inclination angle between the PD and CON groups; however, this metric did differ between the three time points of the stance phase within the PD group. The COM - COP inclination angle increased at FDS and SDS as the lower extremities were further away from each other. Whereas the COM - COP inclination angle decreased at MS as the limb approached single leg stance and COM approached the COP. In previous studies examining individuals with balance impairments related to vestibular deficits and hemiparesis, researchers found that those with balance impairments had a significantly reduced COM - COP inclination angle compared to controls [41]. This contrast to the results of our study may be due to the mild disease state in the PD group in our study.

There were notable differences in dynamic measures of balance during the gait cycle preceding a 90-degree turn. The largest differences in MOS between PD and CON appeared to be at SDS in the M/L direction, with the PD group demonstrating larger MOS distances than the control group. This finding was unexpected as a previous study with the same cohort that examined the MOS variables during the 90-degree turn found the most notable difference between groups during MS [20]. The midstance phase of gait has been identified as the most unstable point of the gait cycle, due to the single limb stance stability demands during this phase of gait [10]. The COM or XCOM approaching the COP at MS might be expected as the participant is in single-leg balance. Individuals in the PD group could have increased M/L MOS in a preemptive attempt to avoid loss of balance. A study by Nilsson et al. also identified that persons with PD experience turning hesitations due to fear of falling [42]. The Nilsson et al. finding, in addition to the slower gait velocity seen in the PD group in our study, may support our speculation. The larger differences found at SDS could be attributed to dynamic balance and gait motor planning demands during this gait phase immediately before the turn. These differences at SDS may be one factor contributing to increased fall risk during turns in persons with PD. Further research is warranted to investigate our premise about individuals with PD with a fall history or more advanced disease stage.

We note several limitations in this study. One limitation is that participants in the PD group were not classified into disease stages based on the Hoehn and Yahr scale or by score on the Movement Society-Unified Parkinson's Disease Rating scale, due to the researcher's limitations in training and formal certification in the administration and scoring of these standardized measures. However, a standardized assessment of balance and gait function was assessed with inclusion criteria that reflect a sample of individuals with mild to moderate PD. Participants in this study were community ambulators and did not have a positive fall history. Therefore, our results cannot be generalized across disease stages or to individuals with high fall risk. The small sample size may have contributed to a reduced statistical power, which framed several conservative conclusions, even among the statistically different findings that we presented. Since data were collected during ON medication times, our findings may not accurately represent dynamic balance during OFF medication times or during periods where medications are wearing off in persons with PD. Finally, Terry et al. [43] and Kazanski et al. [44] have suggested that Hof et al's. [19] original work on the margin of stability for static postures may need to be modified for use in dynamic gait, therefore additional work is needed in this area.

5. Conclusion

This study's findings provide some insight into the biomechanical control of dynamic stability during ambulation and turns in persons with Parkinson's disease. The PD group demonstrated significant differences for all spatiotemporal parameters between Pre-turn and Walking conditions. Decreased step and stride length, as well as cadence and walking velocity for the PD group during Pre-turn, compared to Walking, suggest an adaptation in anticipation of a turn.

Our findings regarding the MOS variables demonstrate important differences that reflect dynamic stability changes during Pre-turn in individuals with mild-moderate PD. In the M/L direction, the PD group's COM is further from the BOS (or COP) as compared to healthy controls. The PD group demonstrated significantly increased MOS in the M/L direction at MS and SDS for all three variables and at FDS for U_{MAX} - XCOM. In the A/P direction for U_{MAX} - XCOM the PD group demonstrated significantly increased MOS at FDS. These changes in the MOS metrics may reflect adaptation within the PD group to increase dynamic stability in anticipation of a turn.

Persons with mild to moderate Parkinson's disease demonstrated changes in spatiotemporal parameters and their margin of stability, particularly M/L stability, during ambulation prior to a 90-degree turn. This information may provide insight into possible factors that contribute to fall risk in this population during walking and turning tasks in this population. Rehabilitation clinicians working with persons with PD may want to carefully assess changes in gait and balance control both in preparation for and during the turn while walking.

Further research in this area should include a larger sample size and persons with moderate to severe PD symptoms who have a greater fall risk or freezing of gait during walking mobility tasks.

Note: This project was part of a larger single-session data collection and some methodological details were not provided in this paper but can be found in Alderink et al. [20].

Acknowledgements

We would like to thank the participants of this study, as well as their families, for volunteering their time to contribute to this research.

Conflict of interest

The authors declare no conflict of interest.

Appendix A: Statistical analysis

Marker Labels	Definition	Position on Patient
LFHD	Left front head	Left temple
RFHD	Right front head	Right temple
LBHD	Left back head	Left back of the head (defines the transverse plane of the head, together with the frontal markers)
RBHD	Right back head	Right back of the head (defines the transverse plane of the head, together with the frontal markers)
C7	7th cervical vertebra	On the spinous process of the 7th cervical vertebra
T10	10th thoracic vertebra	On the spinous process of the 10th thoracic vertebra
CLAV	Clavicle	On the jugular notch where the clavicles meet the sternum
STRN	Sternum	On the xiphoid process of the sternum
RBAK	Right back	Anywhere over the right scapula (no equivalent marker on the left side)
LSHO	Left shoulder	On the acromio-clavicular joint
LUPA*	Left upper arm	On the upper lateral 1/3 surface of the left arm (place asymmetrically with RUPA)
LELB	Left elbow	On the lateral epicondyle
LFRM	Left forearm	On the upper lateral 1/3 surface of the left arm (place asymmetrically with RUPA)
LWRA	Left wrist marker A	At the thumb side of a bar attached to a wristband on the posterior of the left wrist, as close to the wrist joint center as possible
LWRB	Left wrist marker B	At the little finger side of a bar attached to a wristband on the posterior of the left wrist, as close to the wrist joint center as possible
LFIN	Left finger	Just proximal to the middle knuckle on the left hand
RSHO	Right shoulder	On the acromio-clavicular joint
RUPA*	Right upper arm	On the upper lateral 1/3 surface of the right arm (place asymmetrically with LUPA)
RELB	Right elbow	On the lateral epicondyle approximating the elbow joint axis
RFRM*	Right forearm	On the lower lateral 1/3 surface of the right forearm (place asymmetrically with LFRM)
RWRA	Right wrist marker A	At the thumb side of a bar attached symmetrically to a wristband on the posterior of the right wrist, as close to the wrist joint center as possible
RWRB	Right wrist marker B	At the little finger side of a bar attached symmetrically to a wristband on the posterior of the right wrist, as close to the wrist joint center as possible
RFIN	Right finger	Just proximal to the middle knuckle on the right hand
SACR*	Sacral	On the skin mid-way between the posterior superior iliac spines (PSI) and positioned to lie in the plane formed by the ASIS and PSI points
LASI	Left ASIS	Left anterior superior iliac spine
RASI	Right ASIS	Right anterior superior iliac spine

Marker Labels	Definition	Position on Patient
LPSI	Left PSIS	Left posterior superior iliac spine (immediately below the sacroiliac joints, at the point where the spine joins the pelvis). This marker is used with the RPSI marker as an alternative to the single SACR marker.
RPSI	Right PSIS	Right posterior superior iliac spine (immediately below the sacroiliac joints, at the point where the spine joins the pelvis). This marker is used with the LPSI marker as an alternative to the single SACR marker.
LTHI	Left thigh	Over the lower lateral 1/3 surface of the left thigh in line with the hip and knee joint centers
LKNE	Left knee	Laterally, on the flexion-extension axis of the left knee
LTIB	Left tibia	Over the lower 1/3 surface of the left shank
LANK	Left ankle	On the lateral malleolus along an imaginary line that passes through the transmalleolar axis
LHEE	Left heel	On the calcaneus at the same height above the plantar surface of the foot as the toe marker
LTOE	Left toe	Over the second metatarsal head, on the mid-foot side of the equinus break between the forefoot and mid-foot
RTHI	Right thigh	Over the upper lateral 1/3 surface of the right thigh
RKNE	Right knee	Laterally, on the flexion-extension axis of the right knee
RTIB	Right tibia	Over the upper 1/3 surface of the right shank
RANK	Right ankle	On the lateral malleolus along an imaginary line that passes through the transmalleolar axis
RHEE	Right heel	On the calcaneus at the same height above the plantar surface of the foot as the toe marker
RTOE	Right toe	Over the second metatarsal head, on the mid-foot side of the equinus break between the forefoot and mid-foot

Note: Marker labels marked with an asterisk (*) are optional; however, using them improved marker tracking during dynamic trials. The model was modified in the present study with the addition of markers placed on the heads of the left and right fifth metatarsals. Additionally, for the subject calibration trial, markers were placed (later removed for walking trials) on the apex of the right and left medial femoral condyles and right and left medial malleoli.

Table A1.
Plug-in gait markers.

Appendix B: Statistical analysis

This is an exploratory study utilizing a relatively small convenience sample of subjects. As is common for biomechanical studies, numerous gait cycles are observed for each subject with a selection of 'good' gait cycles chosen for analysis. Measurements are then made on each gait cycle. Thus the observational unit is a gait cycle. Gait cycles are considered independent but must be nested within subjects to control for subject differences.

The primary purpose of an exploratory study is to identify potential relationships, not prove clinical practice. As such there is a minimal cost of false positives since follow-up studies are anticipated. False negatives however could discourage follow-up work. We chose to use a standard alpha of 0.05 but note 'interesting' results with higher p-values, especially if the estimated effect size is clinically significant.

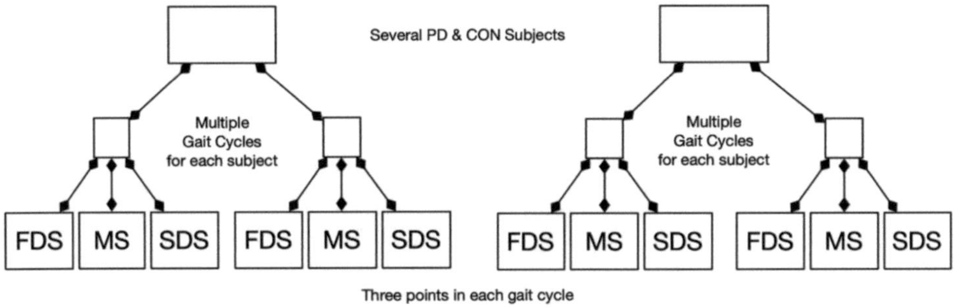

Figure B1.
Margin of stability multilevel mixed model. FDS = first double support; MS = midstance; SDS = second double support.

Prior studies have treated measurements at multiple points in a gait cycle as separate dependent variables. We chose instead to model them as repeated measures in order to get a better overall picture. This approach allowed the identification of surprising differences in the gait cycles based on condition.

The margin of stability (MOS) model chosen is a three-level mixed model **Figure B1**. Level one measurements (primary variables) are taken at three points in the gait cycle called phases (FDS, MS, and SDS), i.e. they are repeated measures within a gait cycle. Gait cycles are level two and random within a subject. Subjects are level three and random with condition (control = CON/Parkinson's = PD) as an attribute. Condition, phase, and their interaction are the fixed effects of interest. Random effects (subject and gait cycle) are used to remove these effects from the model to get appropriate modeling of variance and allow for better detection of effects due to the presence of Parkinson's.

The spatiotemporal model does not have multiple measurements for each gait cycle. Instead, a fixed effect of task (Pre-turn/Walking) is included in the model. The structure is two level with gait cycle at level one within subject level two, both are random. Condition (CON/PD), task (Pre-turn/Walking), and their interaction are the fixed effects of interest.

Several interactions were significant so post hoc t-tests and confidence intervals within levels of condition for MOS variables and Task and Condition for ST variables were performed. Rather than a standardized effect size like Cohen's d, we chose to use the estimated difference since it has direct clinical meaning in this context.

This is a considerably more sophisticated statistical model than often used in these studies but is justified to make the best use of large quantities of data collected on limited subjects. Future statistical work could apply Bayesian statistics and/or consider more extensive multivariate modeling beyond the simple repeated measures used here.

Author details

Gordon Alderink[1*], Cathy Harro[1], Lauren Hickox[2], David W. Zeitler[3], Dorothy Kilvington[1], Rebecca Prevost[1] and Paige Pryson[1]

1 Department of Physical Therapy and Athletic Training, Grand Valley State University, Grand Rapids, MI, USA

2 Department of Mechanical Engineering, Penn State University, University Park, PA, USA

3 Department of Statistics, Grand Valley State University, Allendale, MI, USA

*Address all correspondence to: aldering@gvsu.edu

References

[1] Marras C, Beck J, Bower J, Roberts E, Ritz B, Ross G, et al. Prevalence of Parkinson's disease across North America. npj Parkinson's Disease. 2018;**4**:21. DOI: 10.1038/s41531-018-0058-0

[2] Reich SG, Savitt JM. Parkinson's disease. Medical Clinics of North America. 2019;**103**(2):337-350. DOI: 10.1016/j.mcna.2018.10.014

[3] Park J, Kang Y, Horak F. What is wrong with balance in Parkinson's disease? Journal of Movement Disorders. 2015;**8**(3):109-114. DOI: 10.14802/jmd.15018

[4] Vitorio R, Hasegawa N, Carlson-Kuhta P, Nutt JG, Horak FB, Mancini M, et al. Dual-task costs of quantitative gait parameters while walking and turning in people with Parkinson's disease: Beyond gait speed. Journal of Parkinson's Disease. 2020;**11**(2):653-664. DOI: 10.3233/JPD-202289

[5] Mirelman A, Bonato P, Camicioli R, Ellis TE, Giladi N, Hamilton JL, et al. Gait impairments in Parkinson's disease. Lancet Neurology. 2019;**18**(7):697-708. DOI: 10.1016/S1474-4422(19)30044-4

[6] Center for Disease Control, Web-Based Injury Statistics Query and Reporting System, Inj. Cent. 2020. Available from: https://www.cdc.gov/injury/wisqars/index.html [Accessed: April 8, 2021]

[7] Paul SS, Sherrington C, Canning CG, Fung VSC, Close JCT, Lord SR. The relative contribution of physical and cognitive fall risk factors in people with Parkinson's disease: A large prospective cohort study. Neurorehabilitation and Neural Repair. 2014;**28**(3):282-290. DOI: 10.1177/1545968313508470

[8] Mellone S, Mancini M, King LA, Horak FB, Chiari L. The quality of turning in Parkinson's disease: A compensatory strategy to prevent postural instability? Journal of NeuroEngineering and Rehabilitation. 2016;**13**:39. DOI: 10.1186/s12984-016-0147-4

[9] Justine M, Manaf H, Sulaiman A, Razi S, Alias HA. Sharp turning and corner turning: Comparison of energy expenditure, gait parameters, and level of fatigue among community-dwelling elderly. BioMed Research International. 2014;**2014**:640321. DOI: 10.1155/2014/640321

[10] Kalilani L, Asgharnejad M, Palokangas T, Durgin T. Comparing the incidence of falls/fractures in Parkinson's disease patients in the US population. PLoS One. 2016;**11**(9):e0161689. DOI: 10.1371/journal.pone.0161689. PMID:27583564; PMCID: PMC5008740

[11] Stuart S, Belluscio V, Quinn JF, Mancini M. Pre-frontal cortical activity during walking and turning is reliable and differentiates across young, older adults and people with Parkinson's disease. Frontiers in Neurology. 2019;**10**:536. DOI: 10.3389/fneur.2019.00536

[12] Crenna P, Carpinella I, Rabuffetti M, Calabrese E, Mazzoleni P, Nemni FRM. The association between impaired turning and normal straight walking in Parkinson's disease. Gait & Posture. 2007;**26**(2):172-178. DOI: 10.1016/j.gaitpost.2007.04.010

[13] Kwon M-S, Kwon Y-R, Park Y-S, Kim J-W. Comparison of gait patterns in

elderly fallers and non-fallers. Technology and Health Care. 2018;**26**(S1):427-436. DOI: 10.3233/THC-174736

[14] Huxham F, Baker R, Morris ME, Iansek R. Footstep adjustments used to turn during walking in Parkinson's disease. Movement Disorders. 2008; **23**(6):817-823. DOI: 10.1002/mds.21932

[15] Creaby MW, Cole MH. Gait characteristics and falls in Parkinson's disease: A systematic review and meta-analysis. Parkinsonism & Related Disorders. 2018;**57**:1-8. DOI: 10.1016/j.parkreldis.2018.07.008

[16] King L, Mancini M, Priest K, Salarian R-d-P AF, Horak F. Do clinical scales of balance reflect turning abnormalities in people with Parkinson's disease? Journal of Neurologic Physical Therapy. 2012; **36**(1):25-31. DOI: 10.1097/NPT.0b013e31824620d1

[17] Franzén E, Paquette C, Gurfinkel VS, Cordo PJ, Nutt JG, Horak FB. Reduced performance in balance, walking and turning tasks is associated with increased neck tone in Parkinson's disease. Experimental Neurology. 2009;**219**(2):430-438. DOI: 10.1016/j.expneurol.2009.06.013

[18] Manaf H, Justine M, Goh H-T. Effects of attentional loadings on gait performance before turning in stroke survivors. PM&R. 2015;**7**(11):1159-1166. DOI: 10.1016/j.pmrj.2015.05.007

[19] Winter D. Human balance and posture control during standing and walking. Gait & Posture. 1995;**3**(4): 193-214. DOI: 10.1016/0966-6362(96)82849-9

[20] Alderink G, Harro C, Hickox L, Zeitler DW, Bourke M, Gosla A, et al. Dynamic measures of balance during a 90° turn in self-selected gait in individuals with Parkinson's disease. Applied Sciences. 2023;**13**(9):5428

[21] Hof AL, Gazendam MGJ, Sinke WE. The condition for dynamic stability. Journal of Biomechanics. 2005;**38**(1):1-8. DOI: 10.1016/j.jbiomech.2004.03.025

[22] He C, Xu R, Zhao M, Guo Y, Jiang S, He F, et al. Dynamic stability and spatiotemporal parameters during turning in healthy young adults. Biomedical Engineering Online. 2018;**17**(1):127. DOI: 10.1186/s12938-018-0558-5

[23] Buurke TJW, Lamoth CJC, van der Woude LHV, Hof AL, den Otter R. Bilateral temporal control determines mediolateral margins of stability in symmetric and asymmetric human walking. Scientific Reports. 2019;**9**(1): 12494. DOI: 10.1038/s41598-019-49033-z

[24] Hof AL. The 'extrapolated center of mass' concept suggests a simple control of balance in walking. Human Movement Science. 2008;**27**(1):112-125. DOI: 10.1016/j.humov.2007.08.003

[25] Hof AL, van Bockel RM, Schoppen T, Postema K. Control of lateral balance in walking: Experimental findings in normal subjects and above-knee amputees. Gait & Posture. 2007;**25**(2): 250-258. DOI: 10.1016/j.gaitpost.2006.04.013

[26] Hof AL, Vermerris SM, Gjaltema WA. Balance responses to lateral perturbations in human treadmill walking. The Journal of Experimental Biology. 2010;**213**(15):2655-2664. DOI: 10.1242/jeb.042572

[27] Vlutters M, van Asseldonk EHF, van der Kooij H. Center of mass velocity-

based predictions in balance recovery following pelvis perturbations during human walking. The Journal of Experimental Biology. 2016;**219**(Pt10): 1514-1523. DOI: 10.1242/jeb.129338

[28] Qutubuddin AA, Pegg PO, Cifu DX, Brown R, McNamee S, Carne W. Validating the berg balance scale for patients with Parkinson's disease: A key to rehabilitation evaluation. Archives of Physical Medicine and Rehabilitation. 2005;**86**(4):789. DOI: 10.1016/j.apmr.2004.11.005

[29] Landers MR, Backlund A, Davenport J, Fortune J, Schuerman S, Altenburger P. Postural instability in idiopathic Parkinson's disease: Discriminating fallers from non-fallers based on standardized clinical measures. Journal of Neurologic Physical Therapy. 2008;**32**(2):56-61. DOI: 10.1097/NPT.0b013e3181761330

[30] Hoops S, Nazem S, Siderowf AD, Duda JE, Xie SX, Stern MB, et al. Validity of the MoCA and MMSE in the detection of MCI and dementia in Parkinson's disease. Neurology. 2009;**73**(21):1738-1745. DOI: 10.1212/WNL.0b013e3181c34b47

[31] Giladi N, Tal J, Azulay T, Rascol O, Brooks DJ, Melamed E, et al. Validation of the freezing of gait questionnaire in patients with Parkinson's disease. Movement Disorders. 2009;**24**(5): 655-661. DOI: 10.1002/mds.21745

[32] Yang W-C, Hsu W-L, Wu R-M, Lu T-W, Lin K-H. Motion analysis of axial rotation and gait stability during turning in people with Parkinson's disease. Gait & Posture. 2016;**44**:83-88. DOI: 10.1016/j.gaitpost.2015.10.023

[33] R Core Team. R: A Language and Environment for Statistical Computing. Vienna, Austria: R Foundation for Statistical Computing; 2021. Available from: https://www.R-project.org/

[34] Posit team. RStudio: Integrated Development Environment for R. Boston, MA: Posit Software, PBC; 2023. Available from: http://www.posit.co/

[35] Pinheiro J, Bates D, R Core Team. _nlme: Linear and Nonlinear Mixed Effects Models_. R package version 3. 2023. pp. 1-162, Available from: https://CRAN.R-project.org/package=nlme.

[36] Lenth R. _emmeans: Estimated Marginal Means, aka Least-Squares Means_. R package version 1.8.5. 2023. Available from: https://CRAN.R-project.org/package=emmeans.

[37] Geursen JB, Altena D, Massen CH, Verduin M. A model of the standing man for the description of his dynamic behaviour. Agressologie. 1976;**17**(SPECNO):63-69

[38] Zanardi APJ, da Silva ES, Costa RR, Passos-Monteiro E, dos Santos IO, Kruel LFM, et al. Gait parameters of Parkinson's disease compared with healthy controls: A systematic review and meta-analysis. Scientific Reports. 2021;**11**(1):752. DOI: 10.1038/s41598-020-80768-2

[39] Kalia L, Lang A. Parkinson's disease. The Lancet. 2015;**386**(9996):896-912. DOI: 10.1016/S0140-6736(14)61393-3

[40] Salot P, Patel P, Bhatt T. Reactive balance in individuals with chronic stroke: Biomechanical factors related to perturbation-induced backward falling. Physical Therapy. 2016;**96**(3):338-347. DOI: 10.2522/ptj.20150197

[41] Lee H-J, Chou L-S. Detection of gait instability using the center of mass and center of pressure inclination angles.

Archives of Physical Medicine and Rehabilitation. 2006;**87**(4):569-575. DOI: 10.1016/j.apmr.2005.11.033

[42] Nilsson MH, Hariz GM, Wictorin K, Miller M, Forsgren L, Hagell P. Development and testing of a self administered version of the freezing of gait questionnaire. BMC Neurology. 2010;**10**:85. DOI: 10.1186/1471-2377-10-85

[43] Terry K, Stanley C, Damiano D. A new perspective on the walking margin of stability. Journal of Applied Biomechanics. 2014;**30**(6):737-741. DOI: 10.1123/jab.2014-0089

[44] Kazanski ME, Cusumano JP, Dingwell JB. Rethinking margin of stability: Incorporating step-to-step regulation to resolve the paradox. Journal of Biomechanics. 2022;**144**: 111334. DOI: 10.1016/j.jbiomech.2022.111334

Chapter 5

Community-Based and Everyday Life Gait Analysis: Approach to an Automatic Balance Assessment and Fall Risk Prediction in the Elderly

Britam Arom Gómez Arias, Sebastián Gonzalo Chávez Orellana, Paulina Cecilia Ortega-Bastidas and Pablo Esteban Aqueveque Navarro

Abstract

This chapter discusses the potential of wearable technologies in predicting fall risks among older adults, a demographic susceptible to falls due to age-related walking ability decline. We aimed to explore the feasibility of portable body sensors, mobile apps, and smartwatches for real-time gait analysis in non-clinical, everyday settings. We used classification models like Random Forest, Support Vector Machine with a radial basis function kernel, and Logistic Regression to predict fall risks based on gait parameters. Notably, both Random Forest and Support Vector Machine models demonstrated over 72% accuracy, underscoring the critical role of feature selection and model choice in fall risk prediction. These technologies can enhance older adults' quality of life by predicting fall risks. However, future developments should focus on technologies adapted to non-clinical environments, predictivity, and high-risk group usability. The integration of these features may enable more efficient fall risk assessment systems.

Keywords: gait analysis, fall risk prediction, wearable technology, elderly care, community-based assessment

1. Introduction

Aging is associated with a decline in ambulatory skills as a result of a variety of factors, which have been linked to an increase in the incidence of falls in older adults over 65, being the leading cause of traumatic injury in this population [1, 2].

At least one-third of older adults fall once a year [3, 4]. Alterations in gait and falls are relevant issues in this population, potentially causing limitations in daily life and associating with a range of adverse outcomes, including death [5, 6]. These alterations have a prevalence that progressively increases with age, reaching more than 60% above 80 years old [7].

Importantly, the majority of falls occur during ambulation in the community [1]. Community ambulation has been defined as locomotion outdoors that includes activities necessary for independent living such as visiting the bank, the pharmacy, and the supermarket. Maintaining independent community ambulation can be integral to the quality of life of older adults and their ability to participate socially [6, 8].

Ambulation requires a series of mechanisms that must interact in a coordinated manner for it to be safe and effective [7, 9]. An important determinant is the performance of physical skills such as gait speed and body balance [6, 10]. It has been observed that from 60 years of age, gait speed decreases by about 1% per year [4, 6]. Also, other parameters have been seen to change with age, such as cadence, stride length, angular displacement, joint torque, and power [6, 11].

Commonly, the identification of these variables and parameters in clinical contexts is carried out through scales and tests [12, 13]. Nowadays, due to the subjectivity of these tools [14], the use of instrumentation and/or technologies to increase their predictive value and objectivity has progressively been associated with their use [15, 16].

It is important to note that the majority of these tools are implemented in safe and controlled clinical contexts [14]. However, it has been observed that falls are influenced by various factors, with the environment being a significant determinant in their performance [16, 17]. While the environment alone is not a cause of falls, there is an interaction between the characteristics of the environment that can be dangerous and the intrinsic risk factors of older people [17]. Therefore, restricting evaluations to clinical contexts does not allow for the identification of real issues that may arise according to the different characteristics of the environment and community spaces [14].

The objective of this chapter is to present a novel methodology for gait analysis, concentrating on the application of technological advancements in everyday community settings. We will delve into the current state of the art, highlighting gaps in technology development and its applicability, particularly with respect to gait monitoring initiatives that are based on clinical parameters. Furthermore, we will explore a specific case study that involves an approach based on lower-back accelerometry data. This data, collected from both elderly individuals prone to falls and those not, will be analyzed to evaluate the effectiveness of automated classifiers in predicting fall risks. The overarching aim of this investigation is to expand the scope of gait analysis, bringing it closer to the reality of everyday life for older adults, thereby enhancing its predictive capacity for fall risks.

2. Background and related work

2.1 Methods and technologies for assessing spatiotemporal gait parameters

A substantial body of research has delved into various techniques and devices for assessing crucial spatiotemporal gait parameters such as cadence, step time, and gait asymmetry. The focus of such research often involves practical applications in everyday settings.

The literature reveals wearable body sensors, including accelerometers and gyroscopes, as commonly used tools for this purpose [18–21]. **Figure 1** illustrates some commercially available systems [22–24].

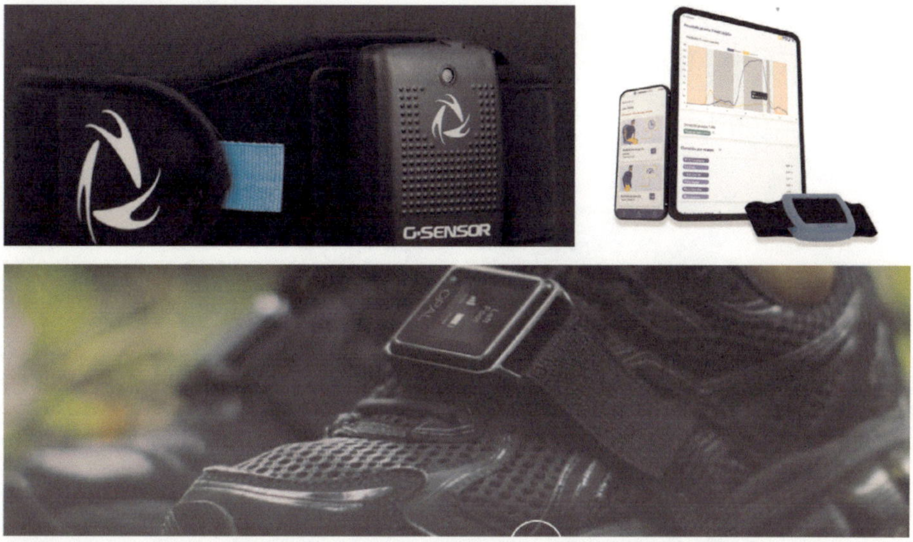

Figure 1.
Commercial systems to assess gait parameters in uncontrolled or community environments. Images from BTS G-walk system website [22], QMUV website [23], and APDM mobility lab website [24].

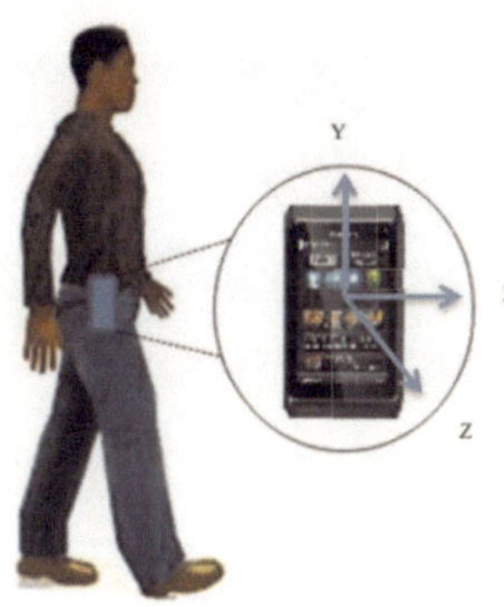

Figure 2.
Mobile application that use in-built sensors to assess gait. Image from ref. [27].

A more controlled approach involves instrumented walkway systems [25], capturing footfall information to derive extensive gait insights.

Mobile technologies have also entered the field, with smartphone applications designed for gait analysis [26–28]. Utilizing the phone's built-in sensors, these applications deliver real-time metrics and feedback (**Figure 2**).

In addition to these, instrumented insoles and shoes [25, 29–32] provide a seamless solution for gait evaluation. Equipped with pressure and motion sensors, these devices offer critical insights into foot loading and movement patterns. **Figure 3** shows a commercial instrumented insole and **Figure 4** illustrates a capacitive instrumented insole to assess gait automatically.

Smartwatches and other commercially available wearables, as mentioned in Refs. [18, 21, 34, 35], offer an attractive, non-intrusive solution. These devices provide continuous monitoring and recording of motion data, allowing for real-time and longitudinal gait analysis.

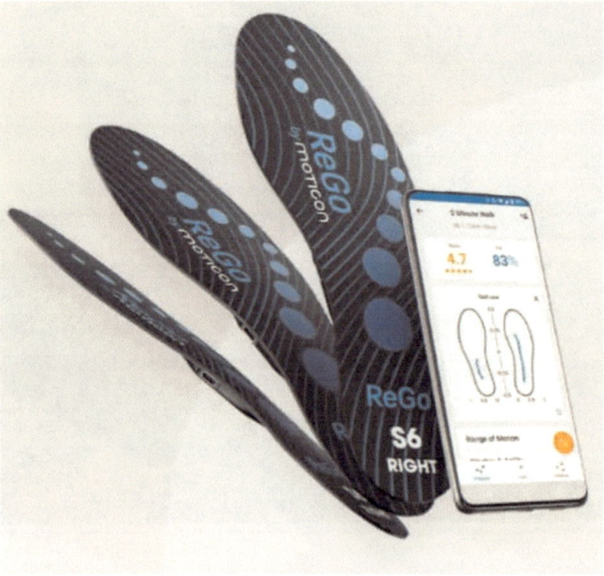

Figure 3.
Commercial systems to assess gait parameters in uncontrolled or community environments. Image from motion sensor insoles system website [33].

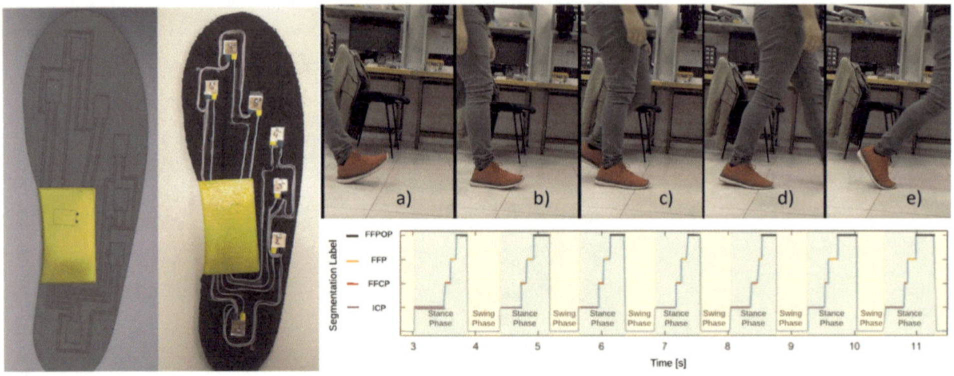

Figure 4.
Capsens SPA instrumented insole system to assess gait automatically. Image from ref. [32].

2.2 Identifying gaps in existing literature

While the variety of gait analysis tools is extensive, certain drawbacks limit their application, particularly in non-clinical, everyday settings. Many of the present technologies require specific conditions for optimum outcomes, restricting their utility for routine use.

A prevalent focus of many tools is to capture existing gait parameters, with little emphasis on predicting future instability or fall risk [18–21]. Additionally, there is a notable scarcity of user-friendly systems that can be easily employed by high-risk groups, such as the elderly [26, 27].

The intricate nature of gait, involving environmental challenges, cognitive distractions, or medication-induced changes, is often overlooked [25, 30, 31].

These observations underline the demand for versatile, easy-to-use gait analysis devices that can be integrated into everyday use. Importantly, these devices should integrate predictive models to spot potential anomalies and elevate fall risk. Catering to these demands could substantially strengthen fall prevention approaches, potentially enhancing the safety and quality of life for high-risk groups.

3. Our approach

Spatiotemporal parameters of gait refer to measures that describe how a person walks in relation to space and time. They are a fundamental aspect of gait analysis, providing valuable insights into a person's locomotion ability.

Estimating the spatiotemporal parameters of gait typically involves identifying specific gait events, which are distinct points in the gait cycle. Here are the primary gait events and how they relate to the estimation of different parameters:

1. Initial Contact (IC) or Heel Strike: This is the moment when the foot first makes contact with the ground.

2. Final Contact (FC) or Toe Off or Pre-swing: The point when the toes leave the ground, initiating the swing phase of the gait cycle.

For more information see **Figure 5**.

While this proposal considers any technological system capable of extracting spatiotemporal parameters from the gait cycle in an uncontrolled or relevant environment for older adults (including smartwatches, smartphones, instrumented shoes/insoles, low back accelerometry, chest accelerometry, in-home video analysis, and more), in this chapter we extracted the necessary parameters for fall risk estimation by analyzing the Physionet "Long Term Movement Monitoring Database" [36]. This

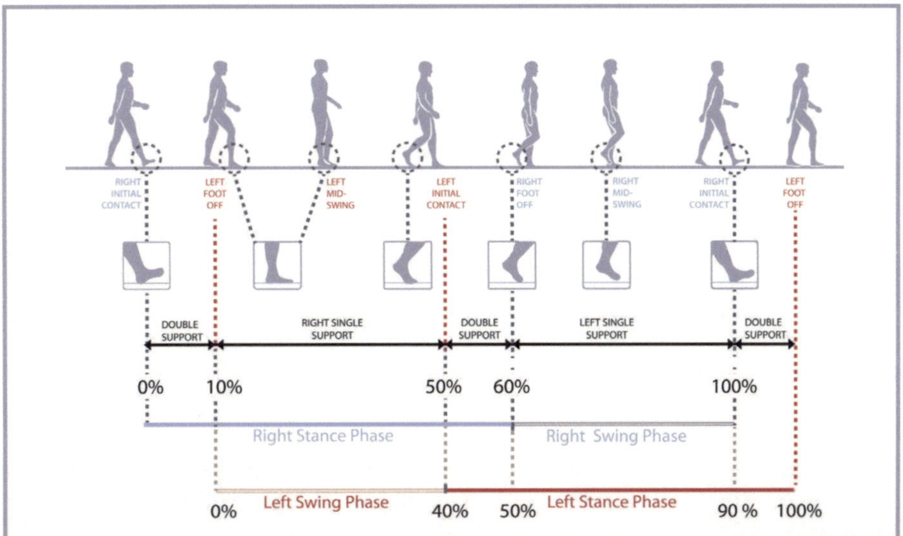

Figure 5.
Gait cycle events and common spatiotemporal parameters [Author's elaboration].

database contains mobility data from 71 elderly individuals (mean age = 78.36±4.71 years; age range 65 to 87 years) during their everyday activities. Within this repository, there are low back accelerometry measurements using a belt with inertial sensors (three-axis accelerometer and three-axis gyroscope) while performing daily life activities, separated into a control group and a fall risk group. These groups were classified based on each volunteer's fall history over the past 12 months as provided by the repository.

The entire process of characterization, feature selection, and training of the classification algorithms was carried out using Python 3.11 with the scikit-learn 1.2.2 package. The development was conducted within the DataSpell 2023.1 Integrated Development Environment (IDE) on a computer running Windows 11 Enterprise (version 22621.963). The system specifications of the computer were as follows: 64-bit operating system, x64-based processor (Intel(R) Xeon(R) CPU E5–2678 v3 @ 2.50GHz, 2.50 GHz), 63.9 GB of installed RAM, and an AMD Radeon RX 5802048SP GPU.

3.1 Feature extraction: Sample analysis

From the Physionet repository, we manually identified and extracted the segments of the measurement that corresponded to independent gait, whether it was in a straight line or otherwise. Of the total test subjects, 69 were evaluated with the method proposed by González et al. [37] for the processing of triaxial lower back accelerometry for the estimations of gait cycle events.

The method consists of processing the signals from each accelerometer axis, which will be mentioned below as: antero-posterior acceleration signal (A–P), vertical acceleration signal (V), and medio-lateral acceleration signal (M–L). This is done to avoid tying the sensor orientation to one and present a general algorithm.

During gait execution, a characteristic cyclic waveform appears from the A–P, V, and M–L accelerometer signals, which can be used to identify the IC and FC events as follows:

1. Filter the A–P signal with an order 11 low-pass FIR filter.

2. Detect two zero-crossings where the area of the filtered A–P is greater than 0.

3. Assess the area validity using Th_{Area} (for this case We used $Th_{Area} = 0.3$).

    a. If valid, continue.

    b. If not valid, return to step 1.

4. Look for the absolute local maximum in A–P. This point corresponds to the IC event.

5. From the IC event to the next zero-crossing on the right of the A–P signal, search for the absolute local minimum in the V signal. This point corresponds to the FC event.

6. Between the identified IC and FC events, look for the absolute maximum and minimum values of the M–L signal.

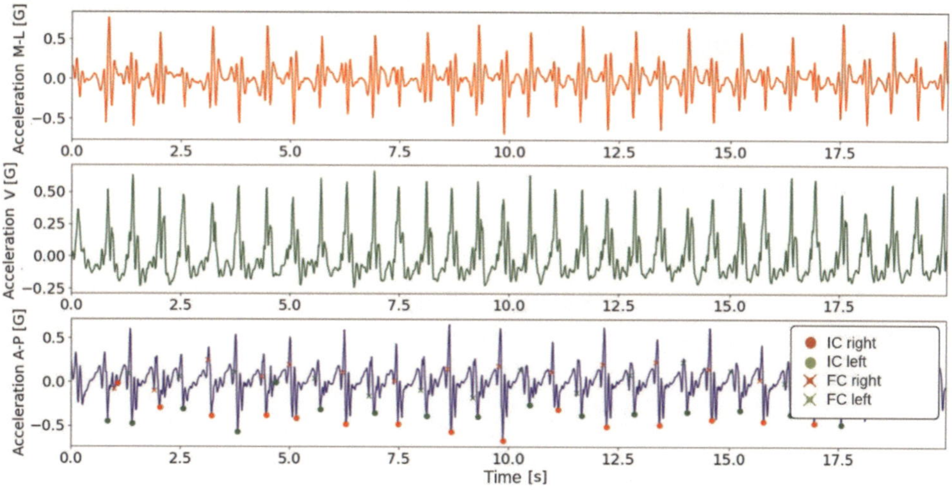

Figure 6.
Gait cycle events identified in a fall-risk subject using a three-axis accelerometer in the lower back [38].

a. If the maximum value is located before the minimum value, the step is right.

b. If the minimum value is located before the maximum value, the step is left.

The result of applying the above algorithm is illustrated in **Figure 5**. Once the gait cycle events are identified, calculations of the spatiotemporal indices characterizing it can be performed (**Figure 6**).

3.2 Gait parameter estimation

To calculate the gait cycle index, it is necessary to identify at least 1 gait cycle for each limb (right and left), which corresponds to the identification of 4 Initial Contact (IC) events and 4 Final Contact (FC) events. If the first detected event corresponds to a right IC, use 1. Otherwise, use 2.

$$cicle_i = \{IC_{R_i}, FC_{L_i}, IC_{L_i}, FC_{R_i}, IC_{R_{i+1}}, FC_{L_{i+1}}, IC_{L_{i+1}}, FC_{R_{i+1}}\} \tag{1}$$

$$cicle_i = \{IC_{L_i}, FC_{R_i}, IC_{R_i}, FC_{L_i}, IC_{L_{i+1}}, FC_{R_{i+1}}, IC_{R_{i+1}}, FC_{L_{i+1}}\} \tag{2}$$

From Eqs. (1) and (2), R corresponds to a right event, L corresponds to a left event and the subscript i is an iterator that goes from 1 to the maximum number of identified cycles. Consider all these events in seconds.

Once at least 1 gait cycle has been obtained, the following temporal indices can be calculated:

- Cadence: Number of steps in 1 minute (see Eq. (3)).

$$Cadence_R = \frac{60}{(IC_{R_{i+1}} - IC_{L_i})}; Cadence_L = \frac{60}{(IC_{L_{i+1}} - IC_{R_i})} \tag{3}$$

- Stride Time: Time it takes to perform a stride, defined as the time between two ipsilateral IC events (see Eq. (4)).

$$StrideTime_R = \left(IC_{R_{i+1}} - IC_{R_i}\right); StrideTime_L = \left(IC_{L_{i+1}} - IC_{L_i}\right) \tag{4}$$

- Step Time: Time it takes to perform a step, defined as the time between two contralateral IC events (see Eq. (5)).

$$StepTime_R = \left(IC_{R_{i+1}} - IC_{L_i}\right); StepTime_L = \left(IC_{L_{i+1}} - IC_{R_i}\right) \tag{5}$$

- Single Support: Total time in which only one foot is in contact with the ground, defined as the time between an ipsilateral FC and IC events (see Eq. (6)).

$$SingleSupport_R = (IC_{L_i} - FC_{L_i}); SingleSupport_L = \left(IC_{R_{i+1}} - FC_{R_i}\right) \tag{6}$$

- Double Support: Time in which both feet are in contact with the ground, defined as the time between contralateral IC and FC events (see Eq. (7)).

$$DoubleSupport_R = (FC_{L_i} - IC_{L_i}); DoubleSupport_L = \left(FC_{R_{i+1}} - IC_{L_i}\right) \tag{7}$$

In addition, parameters can be extracted that immediately identify the general state of the gait:

- Swing Phase: General phase of gait where one foot remains in the air. It is calculated as a percentage of the total of 1 gait cycle (see Eq. (8)).

$$\%Swing_R = \frac{100\left(FC_{R_i} - IC_{R_i}\right)}{IC_{R_{i+1}} - IC_{R_i}}; \%Swing_L = \frac{100\left(FC_{L_{i+1}} - IC_{L_i}\right)}{IC_{L_{i+1}} - IC_{L_i}} \tag{8}$$

- Support Phase: General phase of gait where one foot remains in contact with the ground. It is calculated as a percentage of the total of 1 gait cycle (see Eq. (9)).

$$\%Support_R = \frac{100\left(IC_{R_{i+1}} - FC_{R_i}\right)}{IC_{R_{i+1}} - IC_{R_i}}; \%Support_L = \frac{100\left(IC_{L_{i+1}} - FC_{L_i}\right)}{IC_{L_{i+1}} - IC_{L_i}} \tag{9}$$

- Symmetry Factor: Knowing the duration or percentage of the duration of each phase of the gait (swing and support) for each limb (left and right), a factor can be extracted to observe the symmetry of the gait. Its value lies between 0 and 1. If the factor is close to 0, it indicates asymmetry, where 0 is perfect asymmetry. If it is close to 1, it indicates symmetry, where 1 is perfect symmetry. The calculation is illustrated in Eq. (10).

$$F_{symmetry} = \frac{min\left\{\%Support_R, \%Support_L\right\}}{max\left\{\%Support_R, \%Support_L\right\}} = \frac{min\left\{\%Swing_R, \%Swing_L\right\}}{max\left\{\%Swing_R, \%Swing_L\right\}} \tag{10}$$

Two subjects did not have the data quality necessary to correctly calculate the gait cycle indices, so they were not included in this analysis.

The gait cycle indices that could be calculated from this database using the proposed method were:

- Average left and right step time in seconds.
- Average left and right stride time in seconds.
- Average left and right single support time in seconds.
- Average left and right double support time in seconds.
- Average left and right cadence in steps per minute.
- Average left and right support phase percentage.
- Average left and right swing phase percentage.
- Average gait symmetry factor.
- Walking speed, step speed, and step length could not be estimated using the database as it does not contain demographic information related to the height or leg length of each subject.

3.3 Pre-processing and characterization of spatiotemporal parameters

The database exhibits a balanced distribution, as illustrated in **Figure 7**. Out of the 69 subjects included in the study, 31 are classified as "fallers," while the remaining 38 are categorized as "non-fallers."

In examining the variance of each feature, it becomes evident that the feature "Double Support" demonstrates higher variance compared to its closest characteristic, "Simple Support," as illustrated in **Figure 8**.

As a result, the application of a data scaling method became necessary. By employing the scikit-learn framework, various normalization methods available within its library were assessed, including:

- **MinMaxScaler**: This method scales the features to fit within a specific range, typically between 0 and 1.
- **StandardScaler**: This method transforms the features to have a mean of 0 and a standard deviation of 1.

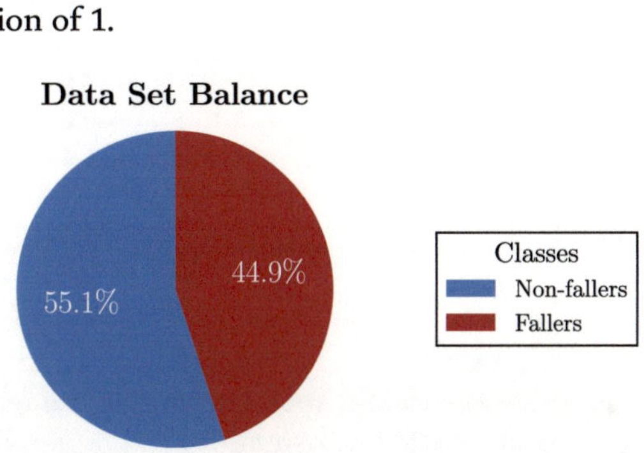

Figure 7.
The pie chart depicts the balanced distribution of subjects in the database categorized by their risk of falling.

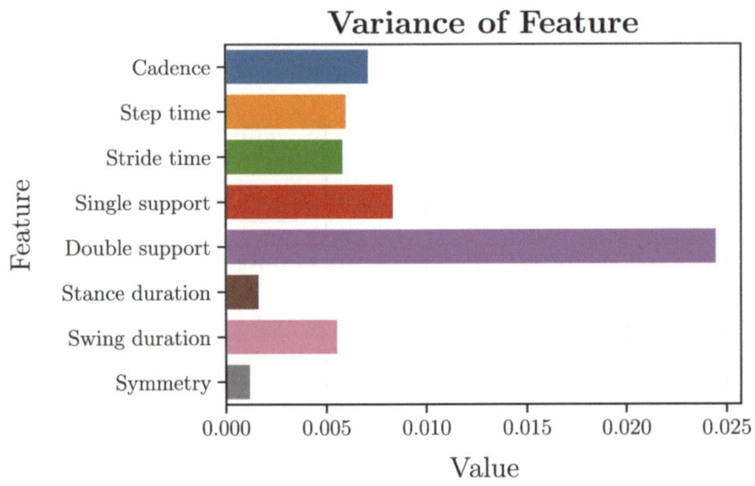

Figure 8.
Representation of the variance of each characteristic using the unscaled data.

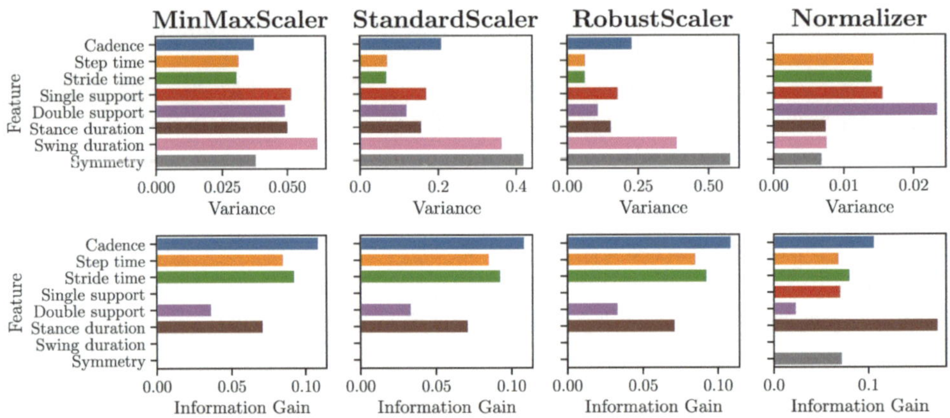

Figure 9.
The sub-figures in the top row show the variance for each feature applying different normalization techniques. The sub-figures in the bottom row show the information gained for each feature according to the type of normalization.

- **RobustScaler**: This approach scales features using statistics that are robust to outliers. Instead of depending on the mean and standard deviation, RobustScaler utilizes the median and interquartile range to scale the data.

- **Normalizer**: This method scales each data sample (row) independently to have a unit norm.

In **Figure 9**, the four methods employed are depicted. The min-max normalization method is observed to generate the least dispersion among the features. Additionally, when considering the estimated mutual information, it is generally apparent that the various types of data scaling do not influence the information contribution from the features to the binary classifier used, with the exception of the Normalizer method.

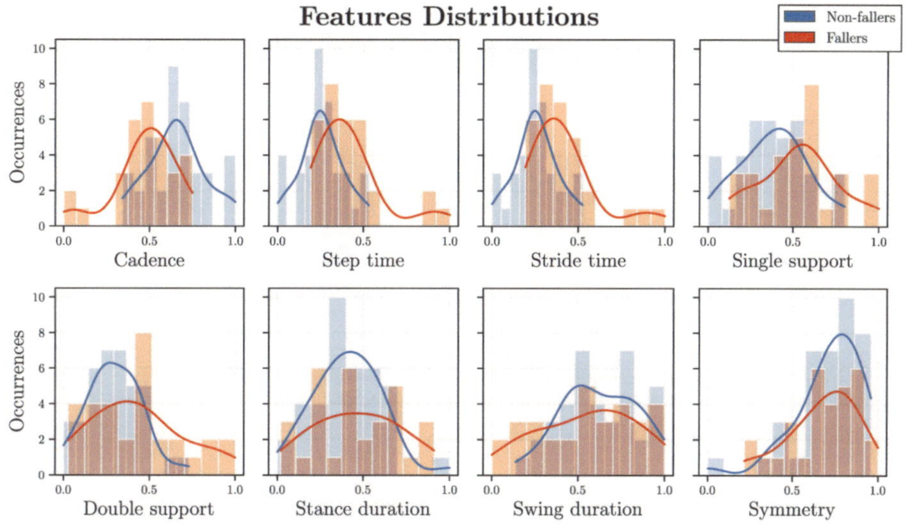

Figure 10.
Histogram for each feature, classified according to class type. The solid lines in the figure represent the kernel density (KDE) for each feature according to the color code of the classifier.

The findings suggest that cadence contributes the most information to the classification problem, followed by stride time, with step time coming in third place.

To determine if there are potential outliers and enhance the interpretability of the data, the distribution of each characteristic was analyzed, as shown in **Figure 10**. From the figure, it can be observed that there are no outliers. To check for normality in the data distribution, the Shapiro-Wilks test was employed with a significance level of 0.05. In seeking to determine the null hypothesis that the data come from a normal distribution, the resulting p-values were evaluated. It was found that characteristics such as cadence, single support, stance duration, and swing duration have a normal distribution, while the rest of the characteristics do not.

In order to investigate the linear dependency between the features, the Pearson correlation coefficient was estimated, and the results were visualized in the correlation matrix, as depicted in **Figure 11**. The findings indicate that cadence exhibits a strong negative correlation with step time, stride time, and single support in comparison with the other features. Conversely, weaker correlations are observed between cadence and double support, symmetry and swing duration, single support and double support, and symmetry with both step time and stride time.

3.4 Training of classification algorithms

To determine the optimal classification model, an exhaustive search was conducted over specific parameter values for an estimator.

For model selection, the GridSearchCV class available in the scikit-learn library was utilized. This class implements an exhaustive search on a grid of hyperparameter values for a given estimator, aiming to identify the best hyperparameter values to maximize the model's performance through cross-validation.

To operationalize this class, a parameter grid was defined in the form of a dictionary, with the hyperparameter names as keys and lists of values to be tested as the

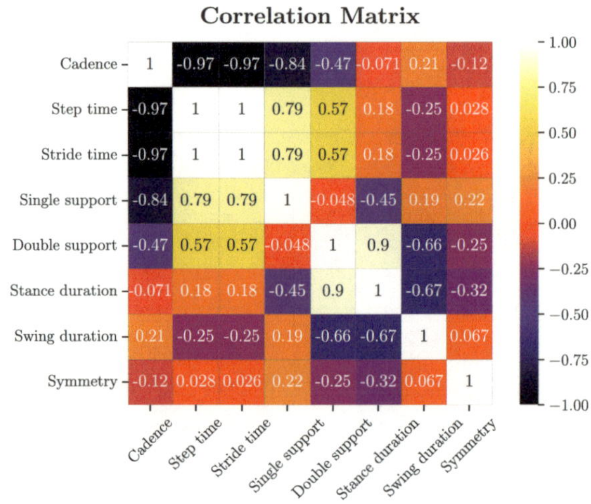

Figure 11.
Matrix of correlations as a heat map. The values represent the Pearson correlation coefficient between the features. A value close to 1 indicates a strong positive connection, a value close to −1 indicates a strong negative connection and a value close to 0 indicates a weak or no connection.

corresponding values. This systematic exploration allowed the evaluation of numerous combinations of hyperparameters, subsequently identifying the optimal configuration for the model.

Regarding the strategy for evaluating the model's performance on the test set, the precision metric was chosen, given the balanced nature of the dataset. This choice is suitable for classification problems when there is an equitable distribution of samples across each class.

For cross-validation, the dataset is configured to be divided into five equal parts. This approach facilitates training and splitting the model into different training and test sets.

The classification models used in this research include:

- K-nearest neighbors (KNN): This machine learning algorithm serves as a nonparametric classification method. Utilizing an instance-based learning approach, it classifies a new data point based on the class labels of its k nearest neighbors in the feature space.

- Logistic regression (LR): A classic classification algorithm, LR is widely used in machine learning for binary classification problems. Unlike linear regression, which predicts continuous values and can produce an infinite number of possible results, LR is tailored for predicting constant or categorical outputs.

- Support vector classification (SVC) with radial basis function (RBF): Support vector machines (SVM) is a supervised learning algorithm that can be applied to classification and regression problems. When used with the radial basis function (RBF) kernel, also known as the Gaussian kernel, SVC can perform nonlinear classification by transforming data into a higher-dimensional space. In the RBF kernel, data points are mapped to an infinite-dimensional space, facilitating the

Model	Accuracy	F1-Score	ROC-AUC	Precision	Recall
RFC[b]	0.7241	0.7500	0.7424	0.8571	0.6667
RFC[w]	0.4138	0.2609	0.3864	0.2500	0.2727
SVC-RBF[b]	0.7241	0.7675	0.7247	0.8125	0.7273
SVC-RBF[w]	0.3448	0.3448	0.3662	0.2778	0.4545
LR[b]	0.7241	0.6923	0.7424	0.6000	0.8182
LR[w]	0.3793	0.5000	0.4646	0.3600	0.8182
KNN[a]	0.6207	0.5926	0.6414	0.5000	0.7273
KNN[w]	0.4318	0.3333	0.5657	0.3077	0.3636

Table 1.
Classification results for different models. The table shows the ranking scores for each model with different combinations of features.

discovery of a hyperplane that effectively separates the classes—even when linear separation in the original feature space is unattainable. The RBF kernel calculates the similarity between two data points in this higher-dimensional space based on their distance.

- Random forest classifier (RFC): Composed of a large number of decision trees, this classifier allows the trees to complement each other independently. When a sample is introduced to the classifier, it is categorized according to the collective voting results of the individual trees.

The results of the classification algorithms using the proposed methodology are presented in **Table 1**.

Next, we will present the details of the training process and feature selection for each classifier.

3.4.1 K-nearest neighbors

For this model, the parameter grid used includes a test list to evaluate the number of neighbors, specifically considering 3, 5, and 7 neighbors. The weights of the function used in the prediction take into account uniform weights and weights that are inversely proportional to their distance. Lastly, a power parameter (p) for the Minkowski metric was considered, which corresponds to using both the Manhattan and Euclidean distances.

The performance of this model relative to others can be observed in **Table 1**. The best results in terms of model performance were achieved when considering the features of swing duration and symmetry, with Accuracy = 62.07%, F1-Score = 59.26%, and ROC-AUC = 64.14%. These results were obtained with the following parameters: a number of neighbors equal to 5, p equal to 1, and weights determined by the inverse of their distance. Conversely, the worst performance was recorded when considering the features of cadence and step time, with Accuracy = 43.18%, F1-Score = 33.33%, and ROC-AUC = 56.57%.

3.4.2 Logistic regression

For this model, no specific values were set in the parameter grid, thereby defaulting to the scikit-learn library's automatic settings.

The optimal combination of features for this model was achieved by considering cadence and step time, resulting in Accuracy = 72.41%, F1-Score = 69.23%, and ROC-AUC = 74.24%. Conversely, the poorest performance was observed when considering swing duration and symmetry, yielding Accuracy = 37.93%, F1-Score = 50.00%, and ROC-AUC = 46.46%.

3.4.3 Support vector classification with RBF

The initial attempts to develop the model involved configuring a parameter grid, where an exhaustive search was conducted using four kernels (linear, sigmoid, rbf, and poly), and the following values for the regularization parameter (C): 2, 4, 3, 5, 6, and 8. Additionally, variations in the kernel coefficient (gamma) were evaluated at 0.1, 0.3, and 0.5, and for the polynomial kernel, the following degrees were examined: 2, 3, 4, 5, 6, and 7. This method facilitated the narrowing of the search for the best classifier, with observations indicating a higher score for the model with the RBF kernel. Consequently, a new exhaustive search was undertaken using a refined parameter grid, employing only the RBF kernel, with C parameters of 1, 2, 4, 3, 5, 6, and 8, and finally, a scaled gamma, the latter being the default parameter used by the class in the scikit-learn library.

The optimal combination of features for this model was found by considering stride time and unique support, resulting in Accuracy = 72.41%, F1-Score = 76.75%, and ROC-AUC = 72.47%. The optimal C parameter was identified as 6, and **Figure 12**

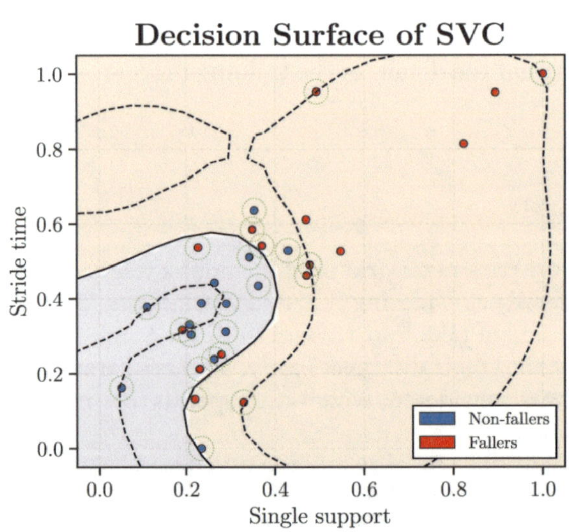

Figure 12.
The decision surface of the SVC model with RBF kernel, with a cost function of 6 and a scaled kernel coefficient. The decision boundary, represented by the solid black line, separates the two classes, with the dashed lines representing the margin boundaries. The support vectors, marked with green dotted line circles, are the crucial data points that define the decision limit. The color map differentiates the two classes, where blue corresponds to the subjects that do not fall and red correspond to those that fall.

illustrates the decision surface of the algorithm. The least effective performance was observed when considering step time, stride time, single support, double support, stance duration, and swing duration, yielding Accuracy = 34.48%, F1-Score = 34.48%, and ROC-AUC = 36.62%, with a C parameter of 8.

3.4.4 Random forest classifier

In training the random forest classifier (RFC), the parameter grid was configured to test various numbers of estimators: 5, 10, 50, 100, and 200. The maximum depth was set at either 5 or 10, and the minimum samples required to split an internal node were set at 2, 3, and 4.

The optimal feature combination for this model was identified by considering single support, double support, stance duration, and swing duration, resulting in an Accuracy of 72.41%, an F1-Score of 75.00%, and an ROC-AUC of 74.24%. The corresponding parameters were found to be a maximum depth of 5, a minimum of 2 samples required to split, and 50 estimators. **Figure 13** illustrates the decision surface for the first tree of the model. Conversely, the worst performance was observed when considering cadence and step time, yielding an Accuracy of 41.38%, an F1-Score of 26.09%, and an ROC-AUC of 38.64%. In this case, the parameters were a maximum depth of 5, a minimum of 3 samples required to split, and 10 estimators.

4. Discussion

Aging brings about a decline in ambulation skills, increasing the incidence of falls in adults over 65 years old. The community represents an environment where the

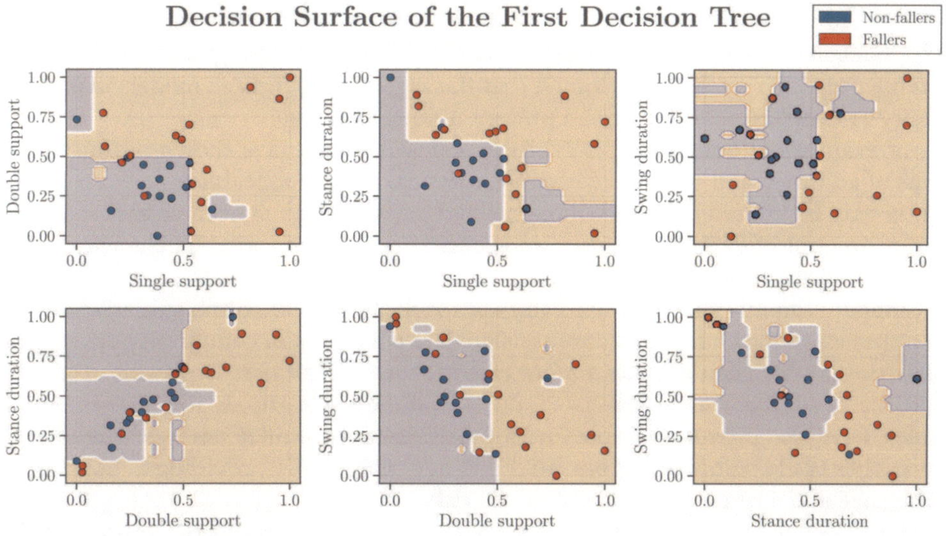

Figure 13.
The decision surfaces of the first decision tree obtained from the RFC, which has been trained with a maximum depth of 5 and 50 estimators. From this tree, the pairs of characteristics used in the decision nodes were obtained, allowing visualization of the decision edges through six subgraphs. Each subplot presents colored training points, where the colors correspond to the non-fallers and fallers classes of the training set. The shaded areas represent the decision regions of the decision tree for each feature pair. The legend provides a key to the classes and colors used in the chart.

majority of these falls occur, making community ambulation critical for the independent and social life of older adults [1]. As such, technology has come to play an essential role in the evaluation of gait cycle variables, such as step time, cadence, and gait symmetry [15, 16].

Portable body sensors, mobile apps, instrumented insoles, and smartwatches have all been used for this purpose [18–35]. Despite the variety of tools, there remain challenges in their application in non-clinical settings and in predicting future instability [18–21].

Body sensors, mobile applications, and smartwatches offer an attractive and non-intrusive solution [22–25, 29–32, 34, 35]. These devices can monitor and record real-time motion data, enabling continuous gait analysis. The availability of commercial systems illustrates the feasibility of these tools in everyday settings. However, the current demand underscores the need for versatile, easy-to-use devices that can be integrated into daily use. These devices should incorporate predictive models to detect potential anomalies and increased fall risk, potentially improving the safety and quality of life for high-risk groups [26, 27].

Based on the results shown in **Table 1**, the various classification models exhibited a wide variation in performance. Random forest (RFC), support vector machine with a radial basis function kernel (SVC-RBF), and logistic regression (LR) models showed outstanding performance in terms of accuracy, F1-Score, ROC-AUC, precision, and recall.

The RFC model with the best features achieved an accuracy of 72.41%, an F1-Score of 75%, and an ROC-AUC of 74.24%. Similarly, the SVC-RBF with the best features demonstrated comparable performance with an accuracy of 72.41%, an F1-Score of 76.75%, and an ROC-AUC of 72.47%. The LR with the best features also showed solid performance with an accuracy of 72.41%, an F1-Score of 69.23%, and an ROC-AUC of 74.24%. In contrast, the k-nearest neighbors (KNN) model had considerably lower performance with an accuracy of 62.07% and an F1-Score of 59.26%.

The diversity in the performance of these models suggests that feature selection and model choice are critical factors to consider when developing fall risk assessment systems. However, it should be noted that the application of these models in a community or everyday life setting can present unique challenges.

For example, although RFC, SVC-RBF, and LR models have shown promise in terms of accuracy and other performance metrics, their practicability in a community setting may be limited by their computational complexity, the need to gather a substantial amount of data for training, and sensitivity to environmental changes and individual variability.

Compared to other technologies and approaches for balance and gait assessment, our approach stands out for its emphasis on data collection in community or everyday life settings, which can provide a more realistic view of fall risk. However, other approaches, such as wearable sensors, mobile phone applications, and instrumented insoles, have also proven valuable for gait assessment in similar settings. These approaches could complement our methodology, potentially providing a more comprehensive and accurate assessment of fall risk.

The evidence presented here suggests several future directions for improving the accuracy and utility of automatic fall risk assessment through gait analysis in community or everyday life settings. There is an urgent need to develop technologies tailoredto non-clinical environments, considering environmental challenges, cognitive distractions, or medication-induced changes. This could include optimizing existing devices, improving human activity recognition algorithms or developing new tools.

5. Conclusions and future directions

This chapter underscores the significant role of technology in assessing gait parameters and predicting fall risks among the elderly in community or everyday settings. The use of portable body sensors, mobile applications, and smartwatches has proven invaluable in monitoring real-time motion data, enabling continuous gait analysis. The accuracy and utility of these technologies have been demonstrated through random forest, support vector machine with a radial basis function kernel, and logistic regression models, which showed commendable performance in their predictive capacities. However, the need for accessible and user-friendly devices is evident, particularly for high-risk groups such as the elderly.

Looking forward, there is immense potential to enhance the utility and accuracy of automatic fall risk assessment through gait analysis in non-clinical settings. The development of technologies tailored to these environments, taking into account factors such as environmental challenges, cognitive distractions, and medication-induced changes, is imperative. As we move forward, a combination of advanced technological tools and improved predictive models can lead to more efficient and effective fall risk assessment, thereby enhancing the quality of life for older adults in community and everyday settings.

Author details

Britam Arom Gómez Arias[1*], Sebastián Gonzalo Chávez Orellana[†],
Paulina Cecilia Ortega-Bastidas[2,3†]
and Pablo Esteban Aqueveque Navarro[†]

1 Faculty of Engineering, Biomedical Engineering, Universidad de Santiago de Chile, Santiago, Chile

2 Kinesiology Department, Faculty of Medicine, Universidad de Concepción, Concepción, Chile

3 Health Sciences PhD Programme, International Doctoral School, Universidad Rey Juan Carlos, Madrid, Spain

4 Faculty of Engineering, Electrical Engineering Department, Universidad de Concepción, Concepción, Chile

*Address all correspondence to: britam.gomez@usach.cl

† These authors contributed equally.

References

[1] Sloot LH, Malheiros S, Truijen S, Saeys W, Mombaur K, Hallemans A, et al. Decline in gait propulsion in older adults over age decades. Gait & Posture. 2021;**90**:475-482

[2] Beck Jepsen D, Robinson K, Ogliari G, Montero-Odasso M, Kamkar N, Ryg J, et al. Predicting falls in older adults: An umbrella review of instruments assessing gait, balance, and functional mobility. BMC Geriatrics. 2022;**22**(1):1-27

[3] Lusardi MM, Fritz S, Middleton A, Allison L, Wingood M, Phillips E, et al. Determining risk of falls in community dwelling older adults: A systematic review and meta-analysis using posttest probability. Journal of Geriatric Physical Therapy (2001). 2017;**40**(1):1

[4] Huijben B, Van Schooten K, Van Dieën J, Pijnappels M. The effect of walking speed on quality of gait in older adults. Alphen aan den Rijn, Netherlands, Philadelphia, United States: Gait & Posture. 2018;**65**:112-116

[5] Marini K, Mahlknecht P, Schorr O, Baumgartner M, De Marzi R, Raccagni C, et al. Associations of gait disorders and recurrent falls in older people: A prospective population-based study. Gerontology. 2022;**68**(10): 1139-1144

[6] Nascimento MDM, Gouveia ÉR, Gouveia BR, Marques A, Martins F, Przednowek K, et al. Associations of gait speed, cadence, gait stability ratio, and body balance with falls in older adults. International Journal of Environmental Research and Public Health. 2022;**19**(21): 13926

[7] Ronthal M. Gait disorders and falls in the elderly. Medical Clinics. 2019;**103**(2): 203-213

[8] Brown CJ, Bradberry C, Howze SG, Hickman L, Ray H, Peel C. Defining community ambulation from the perspective of the older adult. Journal of Geriatric Physical Therapy. 2010;**33**(2): 56-63

[9] Shumway-Cook A, Woollacott MH. Control Motor de la Investigación a la práctica Clnica. Alphen aan den Rijn, Netherlands (global) and Philadelphia, United States (corporate): Wolters Kluwer; 2019

[10] Middleton A, Fritz SL, Lusardi M. Walking speed: The functional vital sign. Journal of Aging and Physical Activity. 2015;**23**(2):314-322

[11] Hollman JH, McDade EM, Petersen RC. Normative spatiotemporal gait parameters in older adults. Gait & Posture. 2011;**34**(1):111-118

[12] Montero-Odasso M, van der Velde N, Martin FC, Petrovic M, Tan MP, Ryg J, et al. World guidelines for falls prevention and management for older adults: A global initiative. Age and Ageing. 2022;**51**(9):afac205

[13] Park S-H. Tools for assessing fall risk in the elderly: A systematic review and meta-analysis. Aging Clinical and Experimental Research. 2018;**30**(1):1-16

[14] Young F, Mason R, Morris RE, Stuart S, Godfrey A. Iot-enabled gait assessment: The next step for habitual monitoring. Sensors. 2023;**23**(8):4100

[15] Ridao-Fernández C, Pinero-Pinto E, Chamorro-Moriana G, et al. Observational gait assessment scales in patients with walking disorders: Systematic review. BioMed Research International. 2019;**2019**:15. Article ID: 2085039

[16] Ortega-Bastidas P, Gómez B, Aqueveque P, Luarte-Martínez S, Cano-de-la Cuerda R. Instrumented timed up and go test (itug)—More than assessing time to predict falls: A systematic review. Sensors. 2023;**23**(7):3426

[17] Pereira CB, Kanashiro AMK. Falls in older adults: A practical approach. Arquivos de Neuro-Psiquiatria. 2022;**80**: 313-323

[18] Werner C, Hezel N, Dongus F, Spielmann J, Mayer J, Becker C, et al. Validity and reliability of the apple health app on iphone for measuring gait parameters in children, adults, and seniors. Scientific Reports. 2023;**13**(1): 5350

[19] Betteridge C, Mobbs RJ, Ho D. Proposed objective scoring algorithm for walking performance, based on relevant gait metrics: The simplified mobility score (smos™)—Observational study. Journal of Orthopaedic Surgery and Research. 2021;**16**(1):1-7

[20] Clay L, Webb M, Hargest C, Adhia DB. Gait quality and velocity influences activity tracker accuracy in individuals post-stroke. Topics in Stroke Rehabilitation. 2019;**26**(6):412-417

[21] Paul S, Tiedemann A, Hassett L, Ramsay E, Kirkham C, Chagpar S, et al. Validity of the fitbit activity tracker for measuring steps in community-dwelling older adults. BMJ Open Sport & Exercise Medicine. 2015;**1**(1):e000013. DOI: 10.1136/bmjsem-2015-000013

[22] Bioingeneering B. G-Walk: Wearable Inertial Sensor for Motion Analysis. [Online]. 2023. Available: https://www.btsbioengineering.com/products/g-walk/

[23] SPA N. Qmuv - el futuro de la rehabilitación. [Online]. 2020. Available: https://qmuv.cl/

[24] Technologies AW. All Day Activity Tracker and More - Apdm Wearable Technologies. [Online]. 2023. Available: https://apdm.com/continuous-recording/

[25] Cramer LA, Wimmer MA, Malloy P, O'Keefe JA, Knowlton CB, Ferrigno C. Validity and reliability of the insole3 instrumented shoe insole for ground reaction force measurement during walking and running. Sensors. 2022; **22**(6):2203

[26] Shema-Shiratzky S, Beer Y, Mor A, Elbaz A. Smartphone-based inertial sensors technology–validation of a new application to measure spatiotemporal gait metrics. Gait & Posture. 2022;**93**: 102-106

[27] Hammoud A, Duchêne J, Abou-Ghaida H, Mottet S, Goujon J-M, Hewson DJ. Validation of a smartphone gait analysis system. In: 6th European Conference of the International Federation for Medical and Biological Engineering: MBEC 2014, 7–11 September 2014. Dubrovnik, Croatia: Springer; 2015. pp. 910-913

[28] Su D, Liu Z, Jiang X, Zhang F, Yu W, Ma H, et al. Simple smartphone-based assessment of gait characteristics in parkinson disease: Validation study. JMIR mHealth and uHealth. 2021;**9**(2): e25451

[29] Ngueleu AM, Blanchette AK, Maltais D, Moffet H, McFadyen BJ, Bouyer L, et al. Validity of instrumented insoles for step counting, posture and activity recognition: A systematic review. Sensors. 2019;**19**(11):2438

[30] el Achkar CM, Lenoble-Hoskovec C, Paraschiv-Ionescu A, Major K, Büla C, Aminian K. Instrumented shoes for activity classification in the elderly. Gait & Posture. 2016;**44**:12-17

[31] Solanki D, Lahiri U. Design of instrumented shoes for gait characterization: A usability study with healthy and post-stroke hemiplegic individuals. Frontiers in Neuroscience. [Internet]. 2018;**12**:459. [Cited 2020 May 2]

[32] Aqueveque P, Germany E, Osorio R, Pastene F. Gait segmentation method using a plantar pressure measurement system with custom-made capacitive sensors. Sensors. 2020;**20**(3):656

[33] Moticon. Moticon Sensor Insoles - Enhancing Sports Health. [Online]. 2023. Available: https://moticon.com/

[34] Huang Y, Xu J, Yu B, Shull PB. Validity of fitbit, jawbone up, nike+ and other wearable devices for level and stair walking. Gait & Posture. 2016;**48**:36-41

[35] Farmer C, van den Berg ME, Vuu S, Barr CJ. A study of the accuracy of the fitbit zip in measuring steps both indoors and outdoors in a mixed rehabilitation population. Clinical Rehabilitation. 2022; **36**(1):125-132

[36] Weiss A, Brozgol M, Dorfman M, Herman T, Shema S, Giladi N, et al. Does the evaluation of gait quality during daily life provide insight into fall risk? A46novel approach using 3-day accelerometer recordings. Neurorehabilitation and Neural Repair. 2013;**27**(8):742-752

[37] González RC, López AM, Rodriguez-Uría J, Alvarez D, Alvarez JC. Real-time gait event detection for normal subjects from lower trunk accelerations. Gait & Posture. 2010;**31**(3):322-325

[38] Aqueveque P, Gomez B, Ortega-Bastidas P, Pena G, Retamal G, Cano-de-la-Cuerda R. Predicting risk of falls in elderly using a single Inertial Measurement Unit on the lower-back by estimating spatio-temporal gait parameters. In: 2022 44th Annual International Conference of the IEEE Engineering in Medicine & Biology Society (EMBC). IEEE. July 2022. pp. 2390-2394